# The Inquisitive Pioneer vol. IV

*The book of At-Home Basic-Materials Science Activities solving with a Slide Rule*

Bryan Purcell

© 2015

This is for my stellar electromagnetic left-hip, Linda W.

# Introduction

The Inquisitive Pioneer Book Philosophy
Bryan Purcell

The following is the Introduction with some modification to the book series, The Inquisitive Pioneer. The basic goals are these that will be discussed here :

1) Science Activities are essential in helping to build conceptual models of science concepts and to foster a growth in Scientific Literacy
2) Science Activities allow for the full expression of the Scientific Method
3) The Scientific Method's core is Data Analysis stemming from Data Measurement
4) Science Activities with basic materials are the best choice for building models of the science concepts under investigation
5) Ordinary Materials brings science and math to everyone – allows the imagination to soar since the Ordinary becomes Extraordinary
6) The Slide Rule is a visual tool that forces the user to employ and sharpen Numeracy skills
7) Between the Slide Rule and the Data Measurement the nearly lost art of Estimation in Numeracy is regained
8) The Inquisitive Pioneer philosophy is to move from the Conceptual, Descriptive, Qualitative to the Mathematical, Relational, Quantitative
9) To be the Inquisitive Pioneer, one must Resolve, Solve, and Evolve! : )

The Inquisitive Pioneer book series are for a thinking person wanting to take a personal journey into fundamental science principles through investigation. The primary person it is for is one who is inquisitive, one who not only likes science and math but prefers that science is explored in a hands-on method using basic, even everyday materials. This science-minded person views the world from the scientific method approach. In life, there is a phenomenon or question, which sparks one's curiosity to do some research and reading. This then leads to posing a measurable hypothesis. From there the science-minded person gathers materials that can readily be at hand or are easy to obtain for use and be used to reach the goals set forth in the hypothesis. Experiments are ran, measurements taken, and numbers analyzed via mathematics to reach conclusions. All along it is the mind of the student of life that assembles these thoughts, these ideas, these numbers measured and mentally constructs a model of what is. Science itself is the story of what is, and is the fabric woven through the whole of the universe and ourselves. We as scientists, uncover the story of science through qualitative descriptions utilizing science nomenclature and the quantitative depictions from the language of mathematics.

A good and personal definition of Science is as follows : Science is the ongoing investigative story of the universe as told by human translation of the descriptions, the interconnections, and the relations of all things in the Universe.

We as teachers, parents, and even as students undertaking the journey of science need to emphasize it is our mind's ability to take in what is, organize the information, and decipher the story of the fabric of the whole of the cosmos from the subatomic to the galactic.
In essence, science is learned by doing. It is useful to have good historical stories of the story of science and the history of those who endeavored before us and see what is, but it is also useful to plant seeds to allow a platform to speak from, a place to explore and experiment and refine our knowledge as necessary.

The Science activities in this book are called hands-on, but many books say this. Here the science explorations go from observing to actually doing the science. All activities have as their core the need to have controlled experiments with measurements and the analysis of the data. This is the heart of true science. Observational and qualitative science are the beginning positions for many things, but this book goes the extra step to all things measured, all measurements used for reasoning and reaching conclusions. We become the researcher, the experimenter, and the mathematician.

This series of books ( this being the 4th volume ) looks to examining the critical needs in the Science classroom and for the student today. These skills are : Scientific Literacy and Numeracy. These two are achieved in the book by helping to increase depth in a given topic area as well as connecting science to our everyday lives through everyday things. These two skills are essential today so as to allow a person to have the greatest array of necessary skills in today's world environmentally, socially, philosophically and economically. The books engage the Scientific Process – which is : To understand science, one is to do science. This means to observe, read, reflect, use logic, to measure, to analyze data, and use information and common sense building block concepts that are the base of the universe in order to reach conclusions.

The philosophy of the book in these Science activities is to encourage an independent thinker and learner. Find the pace you want from the activities. Many activities have a prelude that gives some of the historical and/or mathematical foundations of the exploration. When encountering new ideas, read and reflect on them as well as doing further research. Learn as deep a qualitative, descriptive model of the idea as you can.

When approaching the activity – read it through first and envision it and what should happen first. In setting it up, test it and watch what happens even before measurements. Be willing to redo experiments as needed. Employ the terminology and science vocabulary that comes with the early reading. In the actual Activity, with the measurements, watch the data, its trends, make mental predictions, estimate the answers sought and the conclusions your experiment is pointing to.

There are many activities in the books that make math a part of everyday life and the science activities use basic mostly everyday items so that a student sees that science is not just in a book or a given classroom but is found everywhere.

The materials in the lists are mostly low in cost – they can mostly be found at home right from the start – such as rulers, measuring tapes, tape, measuring cups, string and others perhaps. If not, then they are readily found in inexpensive resource places such as dollar and dollar & more, hardware, and other sorts of stores. In some cases there are a few recommended items that can readily be found online with Amazon and the like, and are often low in cost as well, such as a plastic graduated cylinder or a good lab-grade thermometer. One of the goals of this book was to keep most items low in cost and readily reusable for another experiment in most cases. When we see basic materials as tools, it encourages our imaginations to consider other ideas and other questions. These basic materials are part of the philosophy of learning explored here – where common items become the templates to build ideas upon in our minds, which in turn not only help understanding but are then easily adapted to new situations as needed. It then allows us to explore other frontiers. We each use the regular ordinary items in extraordinary ways. This helps make the concept more concrete, helps the imagination build a model for the idea that stems from the regular items. This in turn acts as the bridge to the mathematical, the quantitative, the relational when it comes to the data and the calculations.

Think, too, to the history of science and math – most of it was forged not only in times of only basic materials but more importantly in the minds of those who gleaned it from nature itself by creating analogies between basic materials used in the lab to explore ideas and phenomena on all scales right outside the lab doors. Here too we as readers of this book and the doers of the science become the inquisitive pioneers like the scientists and mathematicians of yesteryear.

There is a basic reasoning to this book. Besides making science and math easy to do, it brings it to the everyday and brings it home. It allows science to jump off the page of a text book, allows it to leave the classroom, and takes it home and has it spring forth from the mind of the scientifically-minded student. The use of basic materials is to help to create tangible models to think about, recall and learn from in order to grasp key ideas in science. This is not only true for the use of marbles and such to act as masses in one activity or as objects in motion in another activity – these marbles can then be reasoned to be any other needed item of a different scale when needed. This reasoning also to applies to the use the slide rule as the math tool to prompt the mind to take in the data, the numbers, and use them in the relations between the variables as found in the equations and uncover the relations that are indeed found in nature.

Why then, the Slide Rule?

This opening chapter in each of the books is my original essay examining the idea of using a slide rule in the classroom of today ( see ch. I ) – despite all of the available technology. The primary conclusions and goals of the book are quite simple :

1) The Slide Rule though seen by most as an antique is still a valuable tool today in math and science education because it is a tangible tool that strengthens math skills by acting as a bridge between reality and the math that measures it. It enables one to think about and visualize problems; forcing the user to come up with an estimated answer, and the slide rule, itself, yields a reasonable answer that is acceptable in a real-world context.

2) To present a set of basic Science and Math Activities that range in time from middle school to high school using very basic materials, utilizing a recipe of the scientific method as a procedural method to ask a measurable question, use basic materials, engage in a logical procedure, take measurements, and use these measures to do one or more necessary calculations to reach scientifically valid conclusions.

3) Present a brief essay on slide rule history takes a snapshot view of the last 3 centuries ( 1600s-1900s ) where the inclusion of the Slide Rule was a necessary catalyst in the realm of the two major Industrial Revolution periods that affected all nations on Earth and ushered in new sciences, mathematics, innovations and inventions, along with scales of production.

**The slide rule, the tangible bridge for the student, can connect all the core classes and is connecting math to the world of science, carries with it history, famous people and places, and promotes active math engagement by the student. This is used in conjunction with the rationale of the opening of the book where science is tangible and can readily be extended from basic materials which promotes models in the mind to understand concepts and which are explored through carefully crafted experiments by being the Inquisitive Pioneer! Resolve, Solve, Evolve! Explore and Enjoy! – Bryan Purcell**

# Table of Contents

## Table of Contents

# Table of Contents

Important Note of Responsibility :

In the case of all of these Activities the following is to be adhered to – All children and students – you need the support, permission, and help from parents, guardians, and/or teachers to do these Activities. Parents and children alike – read ahead through the whole of the Activity so as to anticipate where there may be areas of concern and important levels of awareness. Always employ safe practices when using any items, such as wearing goggles. Be intelligent, mature, and cautious in the use of items – these are tools and you are using them for specific ends in order to learn. Be aware of all concerns – such as the use of hot or cold items, avoid foods that may have allergies associated with them ( nuts, et al ), sharp objects ( knives, scissors, and other edges ) – and act appropriately, make smart decisions, and be safe. When done with Activities always put away materials, clean up the area, and then address the calculations portion of the Activity.


Master List of Items

This is a nearly comprehensive list of all the items needed for all of the Activities. Recognize that any one Activity will require only a small handful of these and in most cases you might find alternatives that can act as good substitutions. Most items can be found at Dollar and Dollar and More type stores along with hardware stores. A few items left off the list are either very common or in the other case very exclusive to that particular Activity and will probably only be used there. Always consider what is needed and what resources match this safely.

| Slide Rule | Stopwatch | Paper |
|---|---|---|
| Ruler | Marbles | Graph Paper |
| Meter or Yard Stick(s) | Eye dropper | Pencil / Pen / Markers |
| Measuring Tape | Paper Towel | String ( kite ) |
| Sewing Measuring Tape | Popper toy | Styrofoam Cups |
| Measuring Cup Set | Pop Bottle | Plastic Cups |
| Kitchen Mass Scale | Twist Ties | Nuts & Bolts (various sizes) |
| Graduated Cylinder | Electronic Mass Scale | Plastic Wrap |
| Multimeter | Spools | Aluminum Foil |
| Various sprouting seeds | Scissors | Paper Clips |
| Thermometers (lab quality ) | Air or Water powered rocket | Indoor – Outdoor objects to measure |
| Balloons | Stack of Books | Pulleys |
| Rubber Bands | Pennies | Plastic sandwich bags |
| Stairway | Super ball | Water |
| Tension Scale | 7" doll | Large plastic syringe |
| Protractor | Balloon(s) | Skateboard or Roller Skates |
| Goggles | Tape ( regular & duct ) | Gallon container |
| Hole Punch | Straws | Dowel Rods |
| Hooks | Pop Bottle | Chairs |
| Use of Indoor Items : Refrigerator, Freezer, Table, Floor | Pop Corn, Yeast, Baking Soda, Vinegar, Sugar, ice, salt | Batteries for any device used |

## Ch.I
## Why bring the Slide Rule back to the Classroom today?

Using Old Tools to Solve a Modern Problem
The Reintroduction of the Slide Rule to the Classroom
Bryan Purcell

Originally adapted from my article in
Oughtred Society Journal Volume 19, Number 1, Winter, 2010

In the world of education today there is a greater emphasis on trying to find the best way to create new paths to success in learning. Most of these emphasize not only greater capacity in one's knowledge base but also the employment of technology to achieve these ends. The argument goes that technology is the best inroad since it already exists. Why? Basically the quick conclusion comes from these points : children today were born into this era of technology hence have familiarity with it, and technology is the backbone of tools in use today in the workplace.

Though these seem to be valid arguments, they overlook the most critical part of education : the process of learning is not the end product. For example these tools enable a quick solution to a story problem, but the answer is not the goal of education. It is instead the acquisition of skills to enable problem solving and the employment of these skills. This is the fundamental goal of both science and math education today. It is parallel to saying that results speak for themselves in the scientific process, but it is the assessment of the student's good data acquisition and analysis that is the most important.

Because of this, the original assertion by education today, that the use of technology is a necessity, can be considered potentially invalid and there may be other paths to the goal. One of these paths is the use of the slide rule as a math tool, not only to solve problems but also to act as a visual bridge to force the user to engage her mind, tie the concepts of the ideas, the data, and the formulae to the real world and come up with a reasonable answer to the question at hand.

The first area to examine in this dialogue is why such an interest in math and science education? The answer is obvious – the majority of the available jobs, even in a tense economy, are in the areas of math and science – such as jobs in medical fields, engineering, computer technology, and the like. A National Science Board report of 2008 mentions that the needs in these sectors will be triple that of the rest of the job market. Also, the greater one's level of education, the greater one's chances are to find employment as well as to have lifelong higher career earnings. It could even be argued that one's level of opportunity and flexibility in the market and what level to which one ascends are directly related to one's math and science skills.

Compare for example the math skills needed today to operate point-of-service cash registers that have pictures of hamburgers and fries as compared to the real powerbrokers in a business, not the CEO, but the CFO (Chief Financial Officer). These same parallels exist in the realm of investigatory science and research such as found in entry-level assistants and chief engineers.

America's greatest competitive edge has always been found in its creative efforts in the areas of science and math, which launched many endeavors such as the NASA golden decade of the 1960s to go from leaving the planet, stepping into space, and safely undertaking the greatest mission in humanity's history, the journey to the moon.

A second reason for the importance of math and science education is the fact that there is a great deal of academic competition that is not only national but international in scope. and often the skills of American students have been shown to be lacking in math and science. Numerous reports have shown that American students are regularly taking math remedial courses in college; their scores on national tests in these areas are low, and even basic math skills are lacking (The Final Report of the National Mathematics Advisory Panel in 2008 from the U.S. Dept of Education).

To address the issue, the question then arises, "What approach works best?" The argument posed here will explore both the areas of necessary concern and the application of a solution in the form of the slide rule to act as the best tool to affect one's success. The topics to explore in brief are :

1. Estimation and Basic Math Operations,
2. Simple Formula Manipulation and Understanding,
3. Math Areas of Ratios & Fractions & Proportions & Conversions, and finally
4. the Math Topics of Significant Figures & Scientific Notation understanding and use.

The abilities of Estimation and Basic Math Skills go hand in hand. These need to start early and continue to expand as a student progresses through school. The emphasis should be on the student's acting as the computer and not the machine. Reliance on the calculator shifts responsibility from the person to the machine. The answer magically appears in the window of the machine and it does not include estimation at all. The slide rule, however, necessitates that one must practice and use basic math skills and continually employ estimation in order to answer questions.

How many teachers could tell the anecdotal story of the set of students who ask for a calculator before giving an estimate or an answer and with the machine they ask whether they should multiply or divide the numbers? A slide rule cannot be used unless one begins to master these basic skills and employ a mastery of basic numeracy.

Also realize that the slide rule is a natural extension of practices already in place in most elementary school systems. In order to teach numbers, their relative sizes, and concepts like addition and subtraction, the number line is considered the best visual tool. Unlike all of the other colorful tools which have an entertainment value, the number line yields the answer. In fact two of them places alongside each other help the process of learning addition and subtraction.

The same argument is true for the slide rule. Instead of a linear line, it is a logarithmically spaced line and is useful due to the properties of logarithms for visualizing multiplication and division easily. Notice that it would be the next logical step in education; if the number line works, why cannot the slide rule?

To carry this idea further, the next area of concern is formulae and their manipulation. Most formulae in school are linear (such as area of rectangles, miles per gallon, cost per unit item, density, average speed, force, pressure, and even Ohm's Law) and are readily found on a slide rule. An important note : National Standards have these and many more formulae for which students are accountable today.

To illustrate, take distance as a value on the C scale, set it over the time on the D scale and opposite the D index is the average speed. One scale is one variable and the other scale is the other key variable in a formula. One could easily explore relations quickly and effectively. For example, 'how much time at a given speed will it take to cover some given distance?' and the like. Notice the visual link of distance and time needed for a given speed. One has to read across the scale. Conceptualizing the changing of one variable and its effect on another is very easy having this tool.

The next area for exploration primarily is concerned with proportions and conversions. Here the slide rule wins hands down! One easily can solve proportions faster with a slide rule than one can with a calculator. Also, conversions can be treated as a proportion (as can the aforementioned 3-variable functions). Many studies, too, illustrate the lack of skill in converting decimal values to fractions and vice versa. The slide rule accomplishes this visually and shows all related fractions to a given decimal value instantly as contrasted with the ubiquitous calculator.

Finally, in the area of significant figures and scientific notation, the slide rule is again the master math tool. In the real world, we need typically no more than 2 or 3 digits of value in answer. No one measures a room's length and width, and then calculates the area to the 4th or 5th decimal place when buying carpeting or tiles. Also consider the goal here: to acquire problem solving skills. This being the case, does one really gain by multiplying a number with 5 digits with another one? How the slide rule is of value here is that the typical slide rule is accurate to 2-3 digits despite the size of the number.

This last statement is explained by scientific notation, which is of such a value and is directly related to the slide rule. The slide rule has only the numbers 1 to 10 on a typical C scale, yet in reality it has all the numbers that exist! The user must merely put the number in scientific notation. In math, the multiplication of exponents or division of the exponents is readily handled by addition and subtraction.

There are some important final thoughts on these matters where the slide rule is of great interest. First, recognize that no studies of the slide rule have ever been done, not even as compared to the calculator. In the same line the skeptic might add that it is an antique. In a parallel argument, why then do we use measuring tapes still when there are electronic devices for distance, why not just use a microwave instead of an oven and stove (how many cooking shows use the microwave over traditional oven ). And finally, since we primarily use digital clocks, why then continue teaching traditional clock reading?

Ultimately in this idea is the question between the Slide Rule and the Calculator :
Which is best for improving math skills and numeracy? The argument has been presented. It is sound in reasoning and consideration. Finally in this case, to overlook a hypothesis is poor science at best.

Are there other benefits not noted to the slide rule? First, unlike the calculator, the slide rule has an extensive history which can help spark the imagination of presentation and packaging of the ideas about it and its use. It was the most powerful math tool in the history of all handheld devices for 350 years.

Second, the slide rule is directly connected to famous names such as Newton, James Watt, William Oughtred, Joseph Priestly in terms of its construction and use and to those who used it such as Einstein, Hans Bethe, and von Braun as well as including numerous mathematicians, scientists, and engineers.

Third, the slide rule was the first tool outside the human mind used to create most mobile and immobile structures in society such as the Empire State Building, the Golden Gate Bridge, the jet engine, the Panama Canal, and even the Apollo spacecraft.

Third, there are a number of websites that illustrate how to use the slide rule (some in power point format), and no matter the form of slide rule there are no special considerations needed, since the rules for multiplying and dividing do not change despite the style of slide rule. There are even virtual slide rules (see footnotes). Plus there are websites which have slide rule loan programs for a class if a teacher is so interested. Finally one could even download printable scales and have students construct their own slide rule! Imagine making a tool that with the classic 9 scales ( C, D, C1, A, B, L, K, S, T ) rivals the power of a scientific calculator, is personally hand-crafted, and has such a history. With the basic slide rule, the journey of the mind in acquiring problem solving skills and connecting math and science to the universe, can begin.

Web sites

Information: The Oughtred Society : www.oughtred.org
Virtual Slide Rules: Derek's Virtual Slide Rule Gallery :
www.antiquark.com/sliderule/sim
Information, Virtual Slide Rule, Slide Rule power point
presentation on how to use the slide rule, and printable
scales for making a slide rule : www.sliderulemuseum.com
Slide rule plans
Scientific American magazine reference from May 2006
article on slide rules by Cliff Stoll :
www.scientificamerican.com/media/pdf/Slide_rule.pdf
Luis Fernandes, Dept of Electrical & Computer Engineering,
Ryerson University :
http://www.ee.ryerson.ca/~elf/ancient-comp/
sliderule.pdf
Circular Slide Rule by Dr. Charles Kankelborg, Dept of
Physics, Montana State University :
http://solar.physics.montana.edu/kankel/math/csr.html
Math & Science Activities : www.cosmicquestthinker.com

## Data Analysis Math Tool Alternative Consideration

It is agreed upon today that students need to have connections to the ideas they learn and hands-on activities are the first critical step. The next step then is for them to take their measurements and find a way to connect the numbers to concepts. One of the most important goals of science is to **analyze data** to reach mathematical conclusions and find relations between the variables.

**The question then becomes : Is there a different way to examine data?** An interesting approach would be one where the students are not only acting as the scientists taking measurements, but also as the mathematician analyzing their measurements. **The answer is the tool, the 'stick' with numbers on it.**

*What if,* students were to use only low-cost basic tools ( rulers, meter sticks, string, thermometers, small masses, marbles, a personally constructed incline made of meter sticks, stopwatches, mass scales, etc ) for measurable labs. With basic tools the students take measurements themselves and then with the help of the laws of mathematics and through the use of a 'stick' with numbers on it, the students come to discover and find the relations that they can then read about in their texts?

Even in the case of non-measurable labs where the students are merely supplied, straight-forward data, the students can use the very same mathematical 'stick' and find their relations through some graphing and basic computations.

**What 'stick' is this? It is the common slide rule!**

*Why this tool?* The **slide rule** is a tangible and visual bridge connecting numbers to the measured real world. It can be seen as an extension of the use of number lines in their early school journey where they were used for adding and subtracting, only here the slide rule is now used for multiplication and division. The slide rule can also act as a motivation for reasoning and mastery of math.

16

To use a slide rule, one must first estimate answers mentally, know what and why the measured values used are, sequence the mathematical steps of the problem, and understand their place values through scientific notation of both the variables and the answer.

Learning to read the graduations on the slide rule, ( along with learning to use a new tool for calculation ) is useful in itself. *Hence, the student becomes the measuring scientist and the computing mathematician simultaneously once again, like those long ago who used such tools.*

**The most critical present-day problem, then, is to find such a tool.** The references at the end of the article note the International Slide Rule Museum web site, where there is a student-loaner program. For the cost of about $11 per semester, a teacher can be loaned a classroom set of slide rules. There is a power point on how to use a slide rule, along with ideas on its history, and a way to have medals for slide rule competitions as well. Also in the references is a web site, Cosmic Quest Thinker, for suggestions for many science and math activities using slide rules. Each of these has further links to other web sites for virtual slide rules, printable slide rules, publications, even places to assemble one's own classroom set of slide rules and the like.

**Data Analysis with a Slide Rule :**

**In any and all lab situations or even tables of data cases**, the students take recorded ( or given ) data and then merely convert the values into log values of these numbers ( read the log value on the L scale from the data value on the D scale on a slide rule ). Now they proceed to graph a log-log plot of each of the variables, such as :

- log(displacement) vs. log(time) for constant acceleration cases;
- log(period) vs. log(distance) for pendulums or planets ( Kepler's $3^{rd}$ Law );
- log(Force or Intensity) vs. log(distance) for inverse-square laws ( such as gravitational or electrostatic forces or light intensity ), et al ).

In this new log data set on its graph, draw a best fit line through these points, and then find the slope of the line. The slope taken as the ratio of two simplified whole numbers will show the exponential relation between the variables and the exponents involved.

For example, in the case of constant acceleration ( dropped objects or masses on inclines ), the ratio of displacement to time will have a ratio of 2 to 1, hence $d \sim t^2$. This means that the graph is of the form $y = x^2$, which is, indeed, a parabola.

Once the variable relation is found, the slide rule can then be used to then check it as well as explore the relation. Continuing the above example, graph now displacement vs. time-squared as well as the log values of each and again for each draw a best fit line and determine slope. The former determines acceleration while the latter should demonstrate a slope of 1.

In the case of inverse-square laws ( gravitational, electrostatic ), the slope on a slide rule has a ratio of -2 to 1 for Force to Distance. The negative slope is a negative exponent, so it can be seen as $F \sim \frac{1}{d^2}$ . This inverse-square law idea applies to light intensity as well.

Even a situation as complex as Kepler's $3^{rd}$ Law can be examined this way and one finds what Kepler found ( using logarithms, no less ) that the period-squared is proportional to the distance-cubed for a planet ( $P^2 \sim D^3$ ).

**Note that each of these and many more calculations can be done with as simple a tool as a common 9-scale slide rule! This very tool is as powerful as a conventional scientific calculator today.**

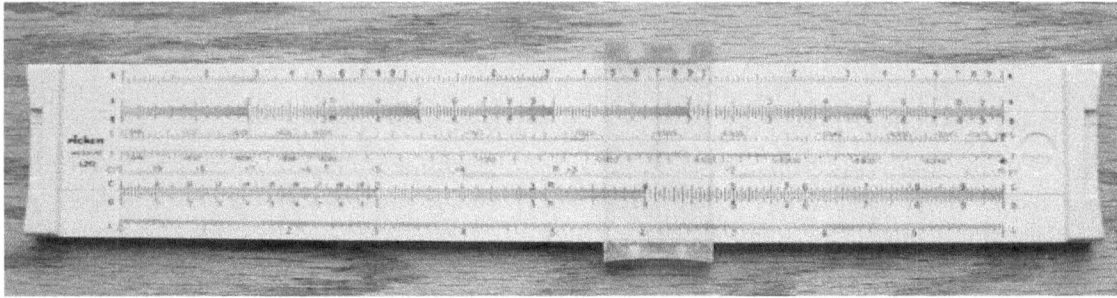

**Other mathematical reasons for the slide rule :** The slide rule can also be used to illustrate the idea of the *laws of logarithms*, such as the product rule for logs where the log of the product of two values is the sum of the logs of each of the values in question. ( log(A*B) = log(A) + log(B) ). Students can compare given values and reach a conclusion here. It is a means to visually conceptualize ideas, such as what happens to variables when one of them changes. **The scales themselves become the variable under consideration.**
Take, for example students given the density of a pure substance. A student places this value on the C scale opposite the left index of the D scale. Now as they read along the C scale, these values represent mass, the numerator of the equation for density ($\rho=\frac{m}{V}$ ), while the adjacent D scale is the corresponding volume value for that given amount of mass so as to always end up back at the material's density! Other types of data analysis can be done this way.
Average Speed is similarly done. Distance is the C scale while Time is the D scale. For any determined average speed, as one reads along the C scale, one has driven farther, hence more time ( D scale ) too. Students can be given data here as well to examine, if preferred.
Because *all similar ratios are set up instantaneously*, this same tool can also be used to easily convert fractions into decimals as well as solve any and all proportions even faster than one can on a calculator. Here tables of information can have blanks to be filled in where students can use the slide rule to find the answers. This can be useful for scaling drawings and maps, calculating changes in recipes, determining cost per unit volume or mass, finding unknown sides of similar polygons, and calculating conversions. **The applications are limited to one's imagination and mathematical skill.**
Notice how this idea extends to a simple activity connecting ideas in math and science in the understanding of the value of $\pi$. Students can measure circumferences and diameters of common circular objects and find the ratio on the slide rule. It will show $\pi$ ( 3.14 ) if done correctly and since all similar ratios are set up, for a given diameter ( or circumference ) they can predict the circumference ( or diameter ). This application applies to any known ratio.
Other scale explorations of the slide rule allow for examination of squares and square-roots ( A & B scales ), cubes and cube-roots ( K scale ), as well as trigonometric relations of sine, tangent, and cosine ( S & T scales ). In combination these scales are all that is needed for virtually all formula in science and math through school. These can all be done as given tables or through measurements depending on the resources and time.
**With this approach, using the slide rule, the goal of having the students do the work and discover the outcome is achieved here.** The goal of data analysis is achieved. When they do a lab and take the measurements, they now take the data and find the relations using math reasoning when using the slide rule. The students here engage in the art and act of discovery through actually doing the math. The students come to find the various relations either through measured or as given data tables. Along the way, they connect the numbers to real-world phenomena.

Also a startling notion develops – *all values measured can be represented as a number between 1 and 10*, as does the slide rule and this promotes the use of scientific notation. Image their surprise when they realize they are holding infinity in one's hands! The use of the slide rule is just an alternative and a way to inspire a path to mathematical reasoning and understanding. Also consider that the slide rule has a sufficient level of precision with 2 or 3 significant figures, which is all that is needed. The tool helps in reinforcing this idea.

Does this mean the end of the computer or calculator? No. *In fact, the calculator and the computer can act now as a follow-up to check the answers. Instead of being the source of the answers, they are the checking system for the student's work as a follow up.*

**What of the use of logarithms and the need to explore them?** This can be done in the science or in math class, if the students are at that level for understanding. Otherwise, letting them know that logarithms are a tool to uncover such relations may be sufficient at this time. As noted here, this idea can be extended to any and all other variable relations they encounter in various science classes as well as math classes.

**This exploration can be a cool math tool adventure.** Students mentally and mathematically examine data themselves to find the answers. They use tools that help them visualize the concepts and make finding answers a personal responsibility and journey. The slide rule promotes math skills acquisition. As an aside, the students can also be introduced to and connected to history through the role of the slide rule. The slide rule was in the hands of numerous scientists, mathematicians, and engineers and used for nearly 350 years ( 1620-1970 ). It has a history of being part of the making of the Panama Canal, the Empire State Building, the Golden Gate Bridge, along with development of the steam engine, the discovery of oxygen, and the determination of the density of the Earth. Both Einstein and von Braun used the same 9-scale model themselves – one from the realm of theoretical physics while the other in the practical realm of applied physics to rocket engineering, where he built the Saturn V, the largest human-made device to leave the Earth carrying aloft Apollo astronauts to the Moon, each carrying a Pickett 600 slide rule.

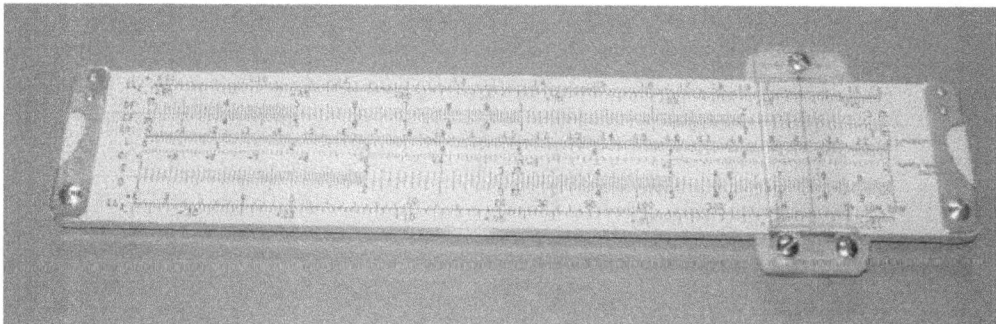

**Resources :**

Slide Rule Loaner Program, Directions for Slide Rule Use, Make your own slide rule :
http://sliderulemuseum.com

Many Classroom Ideas for Slide Rule use :
www.cosmicquestthinker.com

## Ch.III
## Brief overview of the history of the Slide Rule

### Slide Rule History

There were several key historical developments that directly or indirectly fed into the creation of the **Slide Rule**. Europe of the Middle Ages ( 1400 on ) grew in many ways but what affected the Slide Rule were in the areas of the mind, in particular, many new ideas in **math** with the expansion of algebra, geometry and the birth of calculus and logarithms, a growing empirical, objective and **scientific approach** to the world, an increase in speed, volume and efficiency in **communication** with the development of the printing press to allow the flow of new ideas from one place to another, and **transportation**, particularly shipping at first as well as later on in the first Industrial Revolution items such as the steam engine.

All of the aforementioned paved the way to the second Industrial Revolution which was primarily centered on electricity, wireless communication ( radio and television ), construction of massive objects, as well as a faster growing numerical characterization of the universe ( stars, galaxies ). Each of these named in terms of their calculations were connected to logarithms and the primary tool to stem from them, the Slide Rule!

Scientifically-minded people, such as **Galileo,** were the primary catalyst to examine the world in objective empiricism and connect the natural world to the realm of mathematical expression. He is renowned for advancing the scientific method, the basis of kinematics, the foundations of observational astronomy, as well as inventing items. One of his items of interest was the **Sector**. He did not invent it, but improved upon its design which included graduations on movable calipers that could be used for computations involving reciprocals, squares of numbers, and trigonometric formulae. The Sector was used in navigation for the next 300 years.

The foundation of the Slide Rule comes in the early 1600s with **John Napier** ( 1550-1617), who invented the idea of logarithms which he published in **1614** in his book, <u>A Description of the Marvelous Rule of Logarithms</u>. His idea is not the same base as it is commonly used today ( typically 10 or $e$ ) but the idea is the same basis where the value of the logarithm is the exponent of a base number that when the base is raised to this value, it results in the number in question who log is sought.

$$\text{Log}_{\text{Base}}( \text{ Number } ) = \text{Exponent}$$

$$\text{Base}^{(\text{Exponent})} = \text{Number}$$

At first, it seems that complicating numbers to have more calculations to have these results, but it turns out that the properties of logarithms, primarily multiplication, division, raising powers, and taking roots, is actually easier with logarithms. For example :

$$\text{Log}( A * B ) = \text{Log}(A) + \text{Log}(B)$$
$$\text{Log}( A / B ) = \text{Log}(A) - \text{Log}(B)$$
$$\text{Log}( A^N ) = N * \text{Log}(A)$$

Notice that most math expressions, despite complexity, can become nothing more than simple addition, subtraction, or straightforward multiplication. Also it turns out a log-log plot of two variables that one might be considering to have a relation to each other, will have a slope that is a ratio of the powers of the variables in question. With a Log Table, a whole series of multiplications can be done quickly with only addition, and this is not only faster, but can have less error. Tables are created by **Henry Briggs** in **1617**, who employs the use of base 10 in the work : <u>Logarithmorum chilias prima</u>. These come to be called **Briggsian logarithms** and are move within twenty years into Europe and are used the world over for many years. Briggs even gives us the terms describing the components of the logarithm such as 'characteristic' ( the values before the decimal point ) and 'mantissa' ( the values past the decimal point ).

Logarithms are the primary basis of math ( along with Calculus ) to advance the realms of math and science from this time forward and their value cannot be overstated. Logarithms are employed directly and through their visual tool, the Slide Rule to be a part of the process of understanding in science areas such as astronomy, chemistry, physics, as well as the direct application of ideas in navigation, construction engineering for virtually all major immobile and mobile items of the last 3 centuries, and even in the world of business and finance.

The first scientific use of logarithms is by **Johannes Kepler** who employs them ( as well as proving them as being mathematically valid in other works ) in his discovery of his $3^{rd}$ Law of Planetary Motion, relating the period of a planet's revolution with is distance from the Sun. This rule is still employed today for bodies orbiting each other, such as stars and now planets found to be orbiting distant stars as well.

The first physical manifestation of logarithms as a line of numbers appears on what comes to be called the **Gunter Scale** by **Edmund Gunter** ( 1581-1626 ) in **1620**. It is like the Sector, but only a single board about 2 feet in length with various lines inscribed on the wood, including lines of numbers with logarithmic spacing. Using dividers ( aka calipers ) the spacing between the numbers can be spanned so that addition and subtraction can be done, which is really multiplication and division. This is not a Slide Rule, but only one line of what would become a Slide Rule.

Around **1622** ( as best as it can be determined ), **William Oughtred**, a mathematician, aligns two logarithmically-spaced lines so that they can move with respect to each other, hence has what is called the first Slide Rule. Not only does he develop the linear model of the Slide Rule, but also the circular form, he referred to as his Circles of Proportion.

An important change comes from **Seth Partridge** in **1657** where the tool now is given a fixed stock body with a slider, similar to the form the linear Slide Rule is in today.

Even the famous scientist-mathematician, **Sir Isaac Newton** played a part in the history of the Slide Rule. It is noted in his works that while he was working with cubic equation solutions in **1675** he made use of three parallel logarithmic lines and suggested the use of a hairline, what we call a cursor today. This idea was 'rediscovered' a couple of more times through time before it became a regular feature in the later 1800s.

Though more than 50 years since the invention of the Slide Rule, very few people have them, and most are mathematical curiosities. At this point, though the Slide Rule makes a transition into the public sphere primarily through Henry Coggeshall and Thomas Everard.

**Henry Coggeshall** in **1677** develops specialized Slide Rules in the form of the timber and carpenter's rule. In **1683**, **Thomas Everard** uses a specialized one known as a gauging rule that is used to determine the tax on barrels of wines and spirits ( which was actually designed by William Oughtred ). This latter form was used long into the 1800s. He is also the first to name the object what we call it today, the **Slide Rule!** In **1722 John Warner** adds scales for squares and cubes so that these can be determined but so too can square roots and cube roots.

At this point, two things are needed : a more accurate Slide Rule and most importantly communication of its effectiveness and usefulness. This is accomplished by **James Watt** and **James Boulton**, of Industrial Revolution improved steam engine design and development which was assisted by the Slide Rule. These two create what came to be called the **Soho Slide Rule** in **1790**. With regards to the Slide Rule they sought to promote its use by scientists, engineers, and mathematicians. What could be called Technical Journals of the time, such as William Nicholson's "Nicholson's Journal" actually had what would be called Slide Rule ads today.

Innovations to the Slide Rule and accolades for it occur at this time. In Nicholson's Journal, the creation of the folded scale, what is now CF, is noted. In **1797**, **Henry Wollaston** puts in an inverted scale, today known as C1 ( which was reintroduced after 1900 ), and in **1815 Peter Roget** crafts the idea of a log log slide rule, which would not be used for more than 50 years later and then carry on to the 1900s as the central math tool for higher-end engineering. **Joseph Priestley**, the discoverer of oxygen notes in his work that he used a slide rule in his calculations.

Greater standardization comes from **Amendee Mannheim** in **1851** who arranges 4 scales, two double scales, A & B, and two single scales, C & D in a manner that still carries his name for this style, Mannheim arrangement. He also includes a cursor which becomes a more regular feature. Mannheim was in the military where the Slide Rule could be used. The Slide Rule was used a great deal by many nations in both World War I and II and is used in navigation and artillery range determination.

In **1891**, **William Cox** patents what is known as the duplex Slide Rule model with scales on both sides of the tool and are able to be read in conjunction with each other from the cursor placement since all of the scales were based on the C & D scale placement. This patent is held by K&E for some 40 years.

Slide Rules were primarily wood bodies with marks engraved directly on them. In 1886 Dennert and Pape and Faber take wood bodies and place laminated celluloid marked scales on them, which becomes the standard physical form for most of the next century and is one of the most common Slide Rule. In the early 1900s, the Sun Hemmi company is formed and chooses another material for Slide Rules, namely bamboo. It becomes the most common material for Slide Rules in the Pacific Rim arena of Slide Rule production and has great presence in American and European markets as well. In the mid 1900s, with the development of plastic, it became a cost-effective material to mass produce Slide Rules from, especially by the 1960s. Also from the mid 1900s, aluminum becomes a cost-effective tool and is greatly used by Pickett for its Slide Rules.

Though the linear form is the dominant type, there are varieties of them, such as pocket-sized ( 5-6 inches ), regular ( 10 inches ), and 20 inch models. They could also be one-sided ( box-type ) or duplex. The number of scales could range from 2 to over 30! Also in the family of slide rules are Circular Slide Rules, which can be metal with glass faces, such as the Fowler line, plastic, or metal. They too can be single-sided or double-sided as well as varying in size from a 1 inch disk to over 4 inches in diameter. Other varieties include the helical-scale design such as on the Otis King or the Fuller Calculator, where scales can be measured in feet.

The varieties of types, sizes, and materials with the large number of international companies involved in their production ( Nestler, Faber-Castell, K&E, Pickett, Sun Hemmi, Fowler, to name but a few ) plus their changes and filed patents for their form, ( there were only 40 models noted in the span of 1625 to 1800, yet 1800 to 1899 saw the introduction of 250 more types ) illustrate their importance and wide-spread use through time and history.

Also included is how the Slide Rule was developed in Japan in the first place. In the late 1800s, 1894 Japanese ambassador **T. Kondoue** and industrialist/scientist **Dr. R. Hirota** toured Europe during the 2$^{nd}$ Industrial Revolution and one of the key ideas they returned to Japan with and developed was the Slide Rule. It comes to be the largest Slide Rule company in the Pacific, that is Sun Hemmi,  and has sales over 1 million in the later days of the Slide Rule in the 1960s. It is **Jiro Hemmi** who uses bamboo as the core material for these Slide Rules.

Companies like Sun Hemmi and Pickett played strong roles in placing Slide Rules in school with demonstration models and promotions for such things. Sun Hemmi also pushed for making the use of the Slide Rule mandatory in the middle schools and have tests for competency in the high school and civil service exams ( Some nations like France had done this in the mid 1800s too ). The Slide Rule became a world-wide tool found in England, France, Germany, Italy, America, Japan, Australia, U.S.S.R., and many others. It is also interesting to note, though, than an estimated that 40 million Slide Rules were produced during the 1900s.

The Slide Rule had an incredible history with its development and use by many famous people, such as **Albert Einstein** and **Wernher Von Braun** who both used the Nestler 23, a 9-scale model and represented two ends of the physics extremes from the theoretical to the engineering practical with the Redstone Rocket. The Slide Rule also has connections to many famous items directly, such as **the Panama Canal, Hoover Dam, Empire State Building, the Golden Gate Bridge, and the jet engine,** et al ). In the realm of avionics there is **Frank Whittle** with the jet engine and the **F-16**, which was probably the last major jet designed using the Slide Rule. The entire **Mercury, Gemini, Apollo** era of NASA was greatly developed by Slide Rule-bearing engineers, often having a 10 inch model handing from a belt. The Apollo astronauts even had, as a sort of back-up tool, a 6-inch Pickett N-600 Slide Rule.

The Slide Rule became not only the tangible tool derived from logarithms but also became the visual tool that connects our minds to the universe at the large as well as small scales while also being the primary first tool to employ in the construction from the mind to the form for  many mobile and immobile structures in its span of time. This tool, the Slide Rule, is special and becomes all the more brilliant by the genius and creativity of the user and the math journeys undertaken in the cosmos. Pick one up sometime and go on an adventure. Fare thee well.

## How does a Slide Rule Work

The Slide Rule is a mathematical tool that enables the user to perform mathematical calculations of a great variety and obtain a reasonable answer. Though it can be used to add and subtract, it is better just to do this oneself. The Slide Rule is best suited for multiplication and division primarily. But with additional Scales the Slide Rule can be used for other things like squaring, taking the square root, cubing, taking the cube root, determining the power or root of a given expression, determining the log of a given number and the inverse of this by finding the number for a known logarithm and in trigonometry can be used to find the values of the sines, cosines, and tangents of given angles along with their inverses where one has the sine of a given angle and needs to find the angle itself. The range of application mostly depends on the user, her math skills, creative but mathematically sound approach to a problem and the number of scales the Slide Rule has to ease the outcome of sought after answer.

The basic parts ( and these are noted for the linear model ) are the following : The top and bottom strips are called Stators and respectively are referred to as the Top Stator and the Bottom Stator. These pieces are also called Stock in some books. The moving piece between them is the Slide. The Moving Cursor is simple called the Cursor or sometimes called the Indicator or Runner while the cursor line is also known as the Hairline ( it was a hair long ago and first suggested by Isaac Newton ).

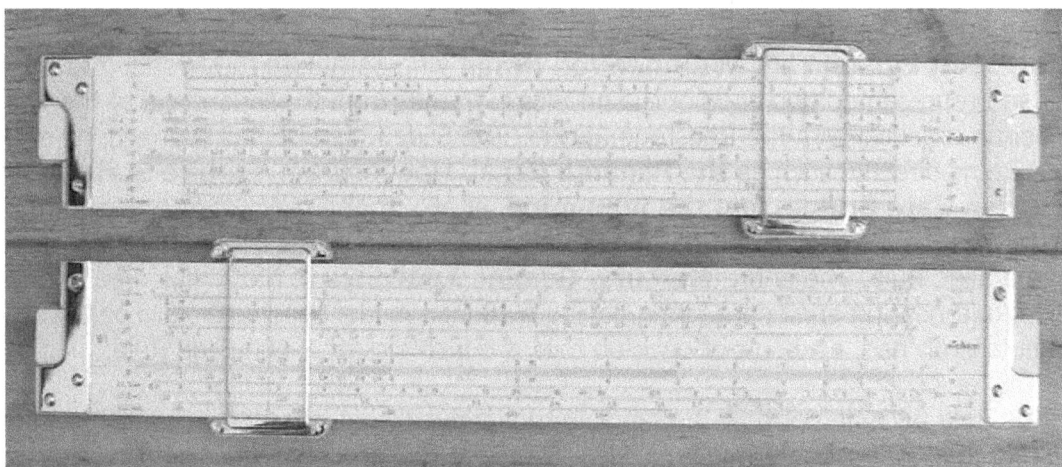

The accuracy of the Slide Rule's answer depends on the Precision of its Scales. Most often the average 10" is effective in its precision of values on its Scales for calculations involving 2 significant digits, but the $3^{rd}$ sig fig can be estimated as well. The number of gradations on the Scales depends on length, so the longer the Scale the larger the number of gradations, hence a greater level of precision can be found. However, it is not as simple as it might first appear. A 20" slide rule merely has a marginally larger number of gradations hence only a slight increase in estimating the 3$^{rd}$ significant figure. To illustrate : between the 9 and right index 1 of the 10" slide rule there are a total of 10 secondary marks, so values like 9.1, 9.2 are easy enough. Between each of these marks in an ever -narrowing gap ( they are spaced logarithmically ) on the 10-inch rule there is a mid-point mark, so this would be 9.15, 9.25, et al. It is easy to conclude that one can estimate values to the nearest 0.01, so values like 9.12, 9.37 can be determined where the 3$^{rd}$ digit is estimated while the first two ( assuming the slide rule is accurately built and the user is mathematically adept ) are reliable. In the case of the 20" rule the same primary and secondary marks exist, but since the scale is longer the tertiary marks now are not 0.5, but instead 0.2. One now has a greater certainty in determining whether the value is indeed 9.13 or 9.14. Notice, however, there is no gain in the number of significant digits though the length of the slide rule has doubled! In general, however, most calculations in the everyday world require only 2 significant digits, so the 10" is sufficient.

The Slide Rule is not a measuring tool, like a ruler, but instead has Scales of Numbers on it that are spaced based on the C & D Scale values which are numbers ( such as 1, 1.2, 2, 3.5, et al ) placed at a distance from the leftmost number '1' value ( aka the Left Index ) that corresponds to the logarithm of that given value and multiplied by the size of the scale ( in a standard rectilinear rule about the size of a ruler this would be 10" or 25 cm ). For example, the number 5 is at a distance from '1' that is log(5)*scale length of the slide rule. This means that all the values are logarithmically spaced from each other on the Scale. What value does having this spacing of values have?

To explore why the numbers on the key base scales ( C & D ) are logarithmically spaced, first we need to explore logarithms themselves. The reason that it makes the Slide Rule a math tool for multiplication and division comes from the properties of Logarithms, which are :

Log(A*B) = Log(A) + Log(B)

Log(A/B) = Log(A) – Log(B)

What this means is this : The log of the product of any two values is merely the sum of the logs of each of the values independently. The log of the division of any two values is the difference of the logs of the respective values as well. What this means is that multiplication can be turned into addition and division can be turned into subtraction. So all we would need then is a table of the log values for any set of values we wish to multiply and we merely add them and find this sum on the table and that would be the product of our values! But the Slide Rule is far easier than this. As noted in the History of the Slide Rule section William Oughtred explored the idea of logarithms and placed the numbers from

1 to 10 at distances from each other that corresponded to the logarithms of the numbers on this line. Knowing the property that the product of any two values is the sum of their logs, all one has to do is combine the distances, hence add them that separate the numbers physically for any given product and the overall total distance will then end up on the number which is the product of the two numbers. Here is a simpler illustration. If I wanted to find the product of 2*3, I merely take the log(2) which equals 0.301 and the log(3) which equals 0.477 and add them together to obtain 0.778 which is the log(6). On the slide rule I merely have to place the starting point, the Index beneath the number in question on another logarithmically-spaced scale – so I place for example the Right Index 1 on the C scale above the 2 on the D scale ( go ahead and do this ). Now read along the C scale to 3 and look back to the D scale and what do you find? 6, of course. You have traveled the distance of 2 on the D scale and the distance of 3 on the C scale – adding these values together is the same as the product of the values since they are logarithmically spaced so we are to the answer of 2*3 or 6. Read more thoroughly the section on Using the Slide Rule to gain greater insight into using the slide rule for basic and more advanced skills as you learn more and do more with it. Notice we do not need to know logarithms themselves, how to derive them, find tables of them, and with a slide rule in hand, we do not have to construct our own math tool to do this.

### Reading the Scale on a Slide Rule :

- All slide rule forms ( linear, circular, et al ) will have the same basic method in reading the scales
- A Scale is the logarithmic-based spaced numbers or a related line to the base scales ( C & D )
- Each mark on a Slide Rule can represent any value as needed through Scientific Notation. So only the values 1 to 10 are needed.
- Watch when doing division or multiplication with scientific notation – see those rules if needed.
- Essentially when values are in scientific notation and are multiplied, then the exponents are added. When values are in scientific notation and they are divided, the divisor ( the denominator ) is subtracted from the quotient ( the numerator ).
- All scales are related to the base scales ( the C and D scales )
- A base scale has the values from 1 to 10, logarithmically spaced.
- However, each mark has a particular reading based on its place on the particular length slide rule.
- The Numbers on the Scale are the Primary Marks. The next major sets of marks are located between them are called Secondary Marks, and when marked, the marks between these are called Tertiary marks.
- When there are 10 marks between any marks, they are tenths (0.1, 1/10)
- When there are 5 spaces between major marks these are two hundredths ( 0.02, 2/100 )

- When a cursor is between any set of corresponding marks, this value becomes the estimated digit and can be done to the level of 0.1 of the major marks involved at best. If very small distances, then it is best to assume only 0.5 of the value between marks.
- The Primary Mark on all scales is the Index which is the number 1. There is a Left one and a Right one on linear slide rules.
- Linear scales ( C & D ) have 2 Indexes while circular ones have 1.
- The Indexes on a linear slide rule are referred to as the Right Index and Left Index.
- The majority of slide rules have capacity for 3 significant figures in their calculations. ( the range is 2 to 4 )
- Best General Rule : Use Scientific Notation in computations
- A, B, C, D, C1, CF, DF, C1F, K, and R scales are all logarithmically spaced values. Some are one time ( C & D ), others ( A & B ) are double, some are triple ( K ), some are reverse ( C1 & C1F ), some are common square roots ( R ), some are folded ( i.e start at a different point ) ( CF & DF – starting at $\pi$ )
- Trigonometric Scales include : S, T, ST
- Log Scales include L, LLN(+ & -) Scales – L is the log value ( usually base 10 ) of a value while the LLN scales are representative of natural log based powers of a C or D scale value to a given exponent ( see LLN scales for more information ).
- Note, unless noted basic math is done with C & D scales
- Scales written in black are from left to right, while in red are right to left
- All basic scales begin with 1 ( except for folded, trig or Log scales ).
- Scales are aligned to read across them.
- When a Slide Rule is a duplex ( two sided ), the cursor can be read for both sides as needed.
- The last digit is estimated in reading a slide rule.
- It is necessary to keep track of the decimal place in calculations.
- Some of the most critical skills in using a Slide Rule are these :
    1. Always mentally keep track of an Estimated Answer ( some sort of range for the answer )
    2. Keep track of the Decimal place in using the slide rule, since it has infinity contained in 1 to 10
    3. Mentally visualize the formula under consideration and project this onto the slide rule so as to keep track of which scales to read

## Multiplication with a Slide Rule :

- Quick summary of rule ( explained step by step below ) : Set one index of the C scale to the multiplicand on the D scale. Next, set the cursor of the runner to the multiplier on the C scale. Finally, read the Answer on the D scale under the cursor hairline.
- [ Note that the answer can be found on either the C or D scale and it depends where one starts and which index is used – it is up to you, but is determined by whether you go off scale – If the calculation is not present, then use the other Index ].

27

- To multiply with a Slide Rule set the Index of one scale ( C ) over the first number to be multiplied on the opposite scale  ( D ). [ Think of this as X from Q = X*Y ]
- Slide the cursor to the second value as it appears on the index scale ( C ). [ This is Y from the equation ]
- Read the answer under the cursor's hairline on the opposite scale ( D ). [ This is Q ]
- Numerical Example :
- Place the Left Index of the C Scale over '2' on the D scale.
- Now read along the C scale to the value of '3'.
- Opposite 3 on C is 6 on the D scale, which is the answer of 2 x 3. This is because you have moved the scales relative to each other both the distance of log(2) and log(3) when added they are the log(6) distance, hence 6 is the answer.
- What is being done is the addition of the distances of the logs of the values in question
- ( Log ( X*Y ) = Log (X ) + Log ( Y ) )
- Though illustrated with C & D scales, note it can be done in reverse ( that is alternate C for D and vice versa ) and this action can also be done on any paired scale-slide combination, such as A & B or if preferred CF & DF.
- With both multiplication and division ( plus other functions ) keep track of the decimal point. Using Scientific Notation is the best choice.
- Scientific Notation Method :
- 
- In Scientific Notation, when multiplying the exponents are added.
- Special Rule : If the slide projects to the Left when performing multiplication, Add One to the Exponent Value for the correct answer.
- ( Note : This is for each calculation, so if done twice, then add two for example )
- ( Note : Also very important these rules apply to only the use of these scales in use, C & D, for example and does not apply to reading across to other scales, such as CF & DF for example )
- In Scientific Notation, when dividing the exponents are subtracted. ( divisor exponent – dividend exponent ).
- One must note, however, whether or not the coefficients when multiplied exceed 10, etc.
- Special Rule :
- Also if the slide in division projects to the right, then the answer from scientific notation exponent needs the subtraction of one from the answer to make it correct. – Note this is for C scale as the Numerator and D scale as Denominator. If the scales used are done in reverse of this, then so too is the rule, hence if it projects to the left in that case then subtract one. ( The idea here is the same as it was for multiplication – the number of times the slide extends to the right, for each time – subtract one from the exponent total).

- 
- Key to calculation is know your decimal placement from products and ratios and using scientific notation it is always one to the right or left of number on the slide rule!

## Division with a Slide Rule :

- Quick summary of rule ( explained step by step below ) : Set the cursor hairline of the runner to the dividend on the D scale. Then slide the divisor on the C scale under the cursor hairline. Finally, read the Answer on the D scale under one index of the C scale.
- [ It is important to recognize that for both multiplication and division the use of the scales can be reversed. Here in step one, C is the numerator while D is the denominator. These roles can be reversed ].
- To divide with a Slide Rule set the divisor value on one scale ( C ) over the dividend value on the opposite scale ( D ).
- Read the answer above the active ( D ) scale index ( 1 ) on the opposite scale ( C ).
- If the answer falls outside the range of the scale ( such as when multiplying or dividing numbers from opposite sides of the scale ) then the other index needs to be used in a linear slide rule.
- ( Note : This does not occur with a circular slide rule ).
- Numerical Example :
- Place '6' on the C scale over '2' on the D scale.
- First note that we are creating a ratio of 6 over 2.
- Next realize that we need to use the Index. In the case of division, always use the Index on the Scale that is the Denominator – here that is the D scale for the example, but it can be the other way if you wish.
- Now notice that the Right Index of D is not opposite any value, so we need to examine the Left Index of D, which is opposite the value of '3' on the C scale, the answer to 6/2.
- What if we wanted 2 divided by 6?
- Here the denominator is the C scale, so again find the useful Index, which is the Right one in this case.
- Since 2 < 6, we expect our answer to be less than 1.
- The right Index of C is opposite 0.333 on the D scale.
- Notice that $3^{rd}$ digit – it comes from the fact that each division between 3.3 and 3.4 is 0.02 and the cursor is in the mid point between 3.32 and 3.34. Finally following the rules of scientific notation and decimal placement, since the slide went to the left and each of the values has an exponent of 0, so 0-0 is 0 and then we subtract one from that and have -1 for our exponent, so our answer of 0.333.
- What is being done is the subtraction of distances of the logs of the two values in question
- ( Log ( X/Y ) = Log ( X ) – Log ( Y ) )
- The best way to consider this operation is think of it as a proportion : the Answer is over 1 while along the scales the Divisor is over the Dividend.

- The Proportion concept is useful for calculations involving conversions and manipulation of 3-variable functions.
- With all slide rules where one scale can move while the other is immobile all answers are present in both multiplication and division as the scales are read. This makes it a parallel calculator – all relations are instantly set up and visual.
- Though the Scientific Notation method works best for values, be wary and if the slide projects to the right and you are reading the Right Index, then in using the Scientific Notation system for the values, subtract one from the answer in all cases. If it projects to the right, then it is alright for division, yet remember that for Multiplication you add one to the total exponent value!

- 
- Alternative Multiplication and Division Method :
- Characteristic Method ( which is akin to the Scientific Notation method, only here the characteristic is the exponent ) :
- For any given set of values, write the characteristic ( the portion in front of the decimal when written as the log of the value in question ( recall the notation : characteristic.mantissa – Note only the characteristic is needed, we are not looking up the mantissa, which, by the way, can be found on the L scale of your slide rule if needed ) –
- Be sure to be wary of positive and negative values in this case!
- Sum up these characteristics.
- Now perform the Multiplication or Division with the Slide Rule as you normally would. That is to say, each number is merely a value from 1 to 10 only. Note that the answer may end up on the range of values before 1 to 10 ( i.e. be 1/10 as large ) or on the range after 1 to 10 and be in the range of 10 to 100.
- For multiplication, each time the slide , with the C Scale, extends past the left index of the D scale, add one ( +1 ) to the Sum of Characteristics.
- For Division, subtract one ( -1 ) from the Sum of Characteristics.
- The revised total is now the characteristic ( i.e. the power of 10 ) for the answer to use with the value showing on your slide rule.
- Return to our 2 x 3 and 6/2 examples for a moment. Each has a Characteristic of 0. In the case of multiplication, it went to the right, so the sum is again 0, while in division the sum is 0, but we subtract one yielding an exponent of -1 from these rules here.


## Combined Operations with a Slide Rule :

- Combined operations are multiple operations in one problem
- In this list there are many examples and notated use of the Slide Rule when it comes to fractions, ratios, proportions, and applications of these ideas.

- 
- Let's say we have the situation for continued products where $Q = A \times B \times C \times$ ---
- Here set the hairline of the cursor indicator at A on the D scale.
- Next move index of C scale under the hairline.
- Next move hairline over B on the C scale.
- Now move the index of C scale under the hairline.
- Next move the hairline over C on the C scale
- Now move the index of C scale under the hairline.
- Continue moving hairline and the index alternately until all the numbers have been set in this case and come to the answer.
- Second scenario :
- If the problem reads ( N x M ) / R then
- Place N ( C scale ) over R ( D scale )
- Slide the cursor along the D scale to M ( on D )
- Find the Answer on the C Scale
- Repeat this process as needed for more than this set up
- Be sure to keep track of Decimal Point as noted above in the Multiplication and Division Rules.
- 
- In essence, combination problems can be seen as proportions which are extensions of ratios, which slide rules are good at.
- For example, any two values over each other is a Ratio or Fraction and once set all similar ratios are automatically established instantaneously. As well opposite the Index of the Divisor is the decimal equivalent of the ratio as well!
- Convert the decimal to a fraction -
- <u>What of converting from a Decimal to a Fraction?</u>
- Place the decimal value on the C Scale over the Right Index of the D Scale and then search along the C and D Scales for a ratio of numbers to represent them.
- Keep in mind this decimal is the percentage value, but you need to multiply in your head by 100 to see it as a percentage.
- For example, 3 on the C Scale over 8 on the D Scale has the Right Index of D under 375 on the C Scale.
- Since 3 < 8, the value is less than one and should be read as 0.375
- What of the Scientific Notation Rules. Each initial value has the exponent 0 and in division, 0 – 0 is 0. But the slide is projecting to the right so we subtract 1 from the result and have -1 for an answer. This means to move the decimal one to the left, and the answer is read 0.375
- As a percentage, multiply by 100 and the answer is 37.5%
- <u>One of the largest uses of fractions is for Sales!</u>
- Place the Price of the Item on the C Scale over the Right Index of the D Scale.
- Read backwards along the D Scale to 9, 8, etc. Each of these is read as N x 10% ( such as 90%, 80% and so on ). Above it is the Price at that percentage!

- 
- Of course, keep in mind, this is not the percentage off, but what you are paying. To see how much is saved, just use the Left Index of D Scale under the Price instead and read to the right instead of left.
- For example, 25% is found at 2.5 and the value above it is 25% of the price. (Of course what you pay is found at 7.5 or 75% instead).
- Sales tax and total cost can be found in a similar manner :
- Place the Cost of the item over the left index of the adjacent scale 9 say C over D as we have been doing )
- Read along the D scale to the sales tax expressed as a decimal value ( this should not be too far, 4% is 0.04, 6% is 0.06, et al )
- The value above on the C Scale is the total cost including sales tax!
- 
- Other Fractions or Ratios in Everyday Life :
- 
- Mpg = miles per gallon
- Take the ratio of miles driven over the number of gallons used sometime to determine just what gas mileage you are getting!
- Mph = miles per hour
- What was your average rate of travel for a trip?
- Take the ratio of the Miles driven to the amount of time ( in hours ) to find the average speed of the trip!
- What if you are some fraction of the distance there and have determined the average speed, then
- The question becomes how much longer 'to grandma's house?'
- Take the remaining distance and divide by the average speed to estimate the number of hours to complete the journey.
- Further, from the mpg calculation you could take the remaining distance divided by the mpg and find the number of gallons needed for the trip in order to decide whether to fill up again or not!


- The fraction as a Slope in Algebra
- A ratio might not be just two numbers, but instead represents the ratio of two differences of numbers :

- $m = \dfrac{\Delta Y}{\Delta X}$

- This is the slope of a linear line, one of the largest topics in algebra.
- The slope is the rate of Rise Over Run. The larger the value, the steeper the line.
- The sign of the slope determines whether the line is moving up or down when examining it from left to right.
- You can consider slope as a 3-variable formula ( see below ).

- The Fraction as a Rate :
- In many practical applications of the fraction it is seen as a **Rate**. The numerator is one value ( distance, gallons, cubic feet of gas, amount of

- growth, temperature change, etc ) while the denominator is some other value ( typically time ).
- The rate could be determined or it may be known in a given problem. If to be determined, the amounts of the other two variables are known
- If the rate is known, then clearly we are missing either the amount in question flowing or the amount of time needed to do this.
- A very good example of a <u>Rate is in Cost per Unit Ounce (Volume)</u>, etc. This is a valuable tool for the slide rule in comparing items when shopping for their comparative costs to find the better deal.
- The flow rate is how water and natural gas consumption is measured and billed. For water and natural gas are Ccf ( 100s of cubic feet ).
- In order to estimate the Cost, one has to merely read the meter at the beginning and the ending times, subtract to arrive at an amount used and multiply this by the cost per unit. This is true for all of the meters : Water, Electrical, and Gas.
- <u>Rates and Ratios are very common in Science :</u>
- In Physics there are many ratios, such as Density ( amount of substance per unit volume ), Pressure ( force per unit area ), Power ( Joules per second ), etc.
- Also in Physics, rates can be speed ( rate of change of distance with respect to time ) or acceleration ( rate of change of velocity with respect to time ) or from Newton's 2$^{nd}$ Law acceleration is the ratio of Net Force to the mass of the object undergoing the net force.
- Chemistry has many ratios such as Molarity ( the number of moles of solute to liters of solution ), Molality ( the number of moles of solute to kilograms of solvent ), Percent by Volume [Mass] ( the volume [mass] of Solute to the volume [mass] of solution ), the Law of Definite Proportions where the Percent by Mass ( is Mass of the Element divided by Mass of the Compound ), as well as the Law of Multiple Proportions ( these could be looked at in the Proportion section obviously ), and so on.
- In Chemistry rates are seen in things such as rate of reaction ( is the negative of the rate of change of reactant to change of time ) plus many more.

- <u>Conversions :</u>

- In Conversions there are two basic methods that can be used with the Slide Rule :
- First, treat the two items as a Fraction that is to be multiplied by the Number in question.
- Typically the ratio of the items is taken where the units one is 'in' are in the denominator, while the units one wants to convert into are in the numerator.

- $\dfrac{\textbf{Units to Convert Into}}{\textbf{Units to Convert From}}$ **or** $\dfrac{X}{Y}$

- This results in this situation :

- 

- Beginning Value in Initial Units* $\frac{\text{Desired Units}}{\text{Initial Units}}$ = Final Value

- $N * \frac{X}{Y} = M$

- The best way to handle this on a slide rule is to :
- Place N on the C Scale over Y on the D Scale on the slide rule
- Read along the D Scale to X, the Desired units and read the answer, M on the C Scale.
- Looking at the prior discussion, it is easy to see that the formula presented can be read as :

- $\frac{M}{N} = \frac{X}{Y}$

- All one has to do is take the ratio of Convert Into Units on the C Scale over the Convert From Units on the D Scale
- Now read along the D Scale to the Beginning Value ( N ) and find the answer above it on the C Scale ( M )
- Use the following Table and look elsewhere for everyday conversions. It is best to keep a small list and memory of these as needed :

- **1 inch = 2.54 cm**
- **12 inches = 1 foot**
- **3 ft = 1 yard**
- **1 mi = 5,280 ft**
- **1 minute = 60 seconds**
- **365 days = 1 year ( rounded )**
- **1 solar day = 24 hours**
- **1 cup = 8 fluid ounces**
- **1 gallon = 4 quarts**
- **1 pound = 16 ounces**
- **1 kilogram = 2.2 pounds**
- **1 ounce = 28.3 grams**
- **1 liter = 1.06 quarts ( rounded )**

- Plus look up whatever you may need !

- For each of the above and any others your find a need for, simply place one value over the other as described above it for use.
- Conversions are needed in many applications :
- One basic unit to another ( inches into feet ),
- Changing one unit type into another ( English to Metric ),
- Currency Conversions, and others.
- Computing Costs

- 
- The reason for conversion depends on the needed outcome. For example feet into yards is commonly used when computing the area of a room in square-yards for carpeting by dividing by 9.
- In the case of painting, the Slide Rule readily calculates the area of a wall with Length times Width, but what of the number of gallons of paint needed?
- Take the total area to be covered and use the Slide Rule to divide by 300 if the walls are unpainted or rough – if smooth and already painted divide by 350 – to determine the number of gallons of paint needed!

- In Math, angles are often converted between Radians and Degrees :

- $180° = \pi$ radians

- In the Sciences, there are numerous conversions not only of the aforementioned units, but also of mixed units :

- Such as km/hr to m/s is very common.
- For that calculation the ratio for it is 3.6/1. Check it yourself!
- In Chemistry there are conversion commonly found in the amount of substances and how it is expressed :
- For example take the number of grams of a substance and divide by its gram molecular weight to determine the number of moles present.
- In Science and Math there are many other conversions depending on the situation at hand.
- In Math in the realm of trigonometry since the C and D Scales are the values for the Sines and Tangents as read from the S and T Scales ( keeping in mind where the decimal falls ), this makes it easy to multiply or divide by the sine or tangent of a value when and where needed.
- Slide the sine value for a given angle read on the C Scale from the S Scale over a given value on the D Scale.
- Read in one way the sine is in the numerator and in the opposite direction it is in the denominator.
- What if you want to find the Sine or Tangent value and are given the sides of a triangle?
- Obviously this is again a ratio : For example :

- $\sin\Theta = \dfrac{\text{Length of Side Opposite}}{\text{Length of Hypotenuse}}$

- Also other angular measures are available and can be used for determining distance or size, such as in the Radian measures :

- $\textbf{Radian measure} = \dfrac{\text{Arc Length of Circle Portion}}{\text{Length of Radius}}$

- If our protractor measures $1/10^{th}$ of a radian, then the apparent size of the object in question is $1/10^{th}$ the measure of the distance from us!

- Another interesting one in Math is the conversion from one base to another in terms of logarithms.
- The standard slide rule has base 10 logs, but let's say you want a number in another base, say 2 or the natural log base e (2.71828*)?
- For any positive numbers, N, A, and B with A $\neq$ 1 and B $\neq$ 1,

- $$\log_A N = \frac{\log_B N}{\log_B A}$$

- From the question the natural log of any value is the ratio of the log base 10 of the number divided by the log base 10 of the natural log value.

- To illustrate, here is a similar example question :
- What is the log base 2 of 6, for example. $\text{Log}_2 (6)$
- Look up reading from the d Scale to the L Scale both the log of 2 and the log of 6. ( 0.301 and .778 respectively )
- Now divide on Scale C and Scale D 778 by 301 - We find it to be 258
- Where does the decimal go?
- Since 6 is much greater than 2, it will have a characteristic ( here 2 and the rest is the mantissa 0.58 ) so the answer is 2.58
- So $2^{2.58} = 6$ ( try it and see, rounded off of course )
- The idea of the ratio, fraction goes on to any and all applications.

- **The Proportion :**

- As noted in the prelude, the proportion is two ratios set equal to each other.

- In the conversions section above, it is easy to see its value in use.
- The basic proportion can be expressed as :

- $$\frac{M}{N} = \frac{X}{Y}$$

- All one has to do is take the ratio of known values - M on the C Scale over N on the D Scale

- Now read along the D Scale to the other known value ( Y ) and find above it on the C Scale the answer ( X )
- It is easy to see that all one needs to know is any 3 of the variables and the $4^{th}$ is the one to find.
- By using the scaled as fractions it is easy to place one value on one scale and the other on the opposing scale.
- Try this for yourself to find that solving a proportion on a Slide Rule is indeed much faster than one can solve one on a Calculator!

- 
- Also here there are no complex rules, like the calculator, such as cross-products –
- In the case of the Slide Rule the natural form is maintained which is the equivalence of two ratios.
- This is an invaluable tool in Algebra for proportions as well as in geometry for any and all similar figures to determine unknown sides!
- Proportions can be used to determine height or distances -  In this case we are using similar triangles.
- For example, hold up a ruler at arm's length to measure the apparent height of a distant object, say a picture on the wall.
- If you read the apparent height, measure your arm's length, ( this is the first ratio )
- now measure the distance to the wall,
- Set the first ratio of your measures to the ratio of the unknown height on the wall to the distance to the wall.
- Then the height of the picture can readily be determined.
- Though simple, this technique is used in surveying regularly and is used in the wilderness for distances across rivers and gorges before the advent of electronic equipment.
- The easiest way to envision it is when you are outside and you as well as a tree or a flag pole casts a shadow on a sunny day.
- The ratio of your shadow length to your height is equal to the shadow length of the flag pole to its actual height.

- $$\frac{\text{Length of your shadow}}{\text{Your Actual Height}} = \frac{\text{Length of flagpole shadow}}{\text{Actual Height of flagpole}}$$

- A more thought out activity involves determining the size of the Moon :

- Use a meter stick and place a small card vertically with a hole punched in it at a distance
- Look along the stick through the hole so that when viewing the full Moon it fully fills the diameter of the hole ( i.e. move the card until you have proper alignment )
- The ratio of the diameter of the hole to the distance that the hole is from your eye equals the actual diameter of the Moon to the Moon's distance from you. ( Here, we assume we know the distance to the Moon ). Hence the diameter of the Moon can be determined!

- Proportions can also be used in changing the scale of a recipe :

- $$\frac{\text{Recipe Requirement for Material}}{\text{Recipe Number of Servings}} = \frac{\text{The amount of Material Needed}}{\text{Number of Servings}}$$

- Here the ratio of how much is needed to the number of servings is set equal to the amount of unknown material and the number of desired servings. The amount needed is readily found.


- **In Math there are many applications :**

- The very nature of the proportion stems directly from geometry and the relations found in similar figures.
- These can be the aforementioned triangles but also includes any similar polygons, such as squares, rectangles, and the like where one is known, the other is partially known and there is a missing side.
- Still other examples are considered :
- *What if one wants to find the circumference of a given diameter of a circle ( or vice versa )?*
- Simply put $\pi$ on the C Scale of the slide rule over the Index on the D Scale. – Note with CF & DF scales this is already done since these scales are set at $\pi$ as their beginning point over the C & D Indexes! See CF & DF scale use
- Now read along the known scale. The C Scale is the Circumference, while the D Scale is the diameter.
- *What about the diameter if the area of the circle is known?*
- Take the Area on the A Scale over p on the B Scale.
- The diameter-squared is found on the A Scale opposite the B Index.
- The diameter is then found below reading from the A Scale to the D Scale to read the square root of the value on A.

- Other interesting things can be done with gauge marks ( special marks on slide rules for conversions, multiplications, et al like $\pi$ ) as well as personally derived values :

- On some Slide Rules there is a mark, c, on the C & D Scales (1.273) which is $\frac{}{\pi}$ and comes from $\mathbf{A} = \pi * \frac{\mathbf{d}^2}{4}$
- Place this C mark or value of the C Scale over the index of the D Scale
- Slide the cursor over the size of the diameter of a considered circle on the D Scale.
- To find the Area of a circle read the answer on the B Scale!
- What if you want to do this by knowing the radius ( of course we could simply multiply by 2 to use the former method, but give your mind a chance to explore the Slide Rule ! )
- Look up the square root of $\pi$ by first finding it on the B Scale and noting its value on the C Scale.
- Slide the square root of $\pi$ value on the C Scale to the Index of the D Scale
- Now read along the D Scale to any desired radius value for a circle.

- The answer in this case is read on the B Scale for the Area of the Circle in question!

- 

- What about the Volume of a Sphere?
- Now place the value 1.61(2) on the D Scale over the index of the C Scale.
- Read along the D Scale, for the radius of a given sphere you are considering.
- The Volume of this sphere is found on the K Scale!
- This comes from $(\frac{4*\pi}{3})^{1/3}$ from the formula $V = \frac{4*\pi*r^3}{3}$
- There are many other problems in Math and Algebra in the area of problem solving:
- For example, if a given material costs so much per ounce, pound, ton, how much will another desired amount cost ?

- $$\frac{\text{Cost}}{\text{Unit Amount}} = \frac{\text{How much does it Cost?}}{\text{Amount Wanted}}$$

- There are numerous ratios of values that can be found or derived from many other references that can be used in proportions to find the answer to many a question one might encounter in a math and or science text:

- $$\frac{\text{Diameter of Circle}}{\text{Side of Inscribed Square}} = \frac{99}{70} = \frac{\text{Diagonal of Square}}{\text{Side of Square}} = \sqrt{2}$$

- Here is a general value for the pressure one feels with depth :

- $$\frac{\text{Pounds per Square Inch}}{\text{Feet of Water}} = \frac{26}{60}$$

- Even more complex proportions can be solved :

- A classic algebra question might read :
- *If it takes 4 people 7 days to accomplish a job, how much time is needed ( assuming the same work rate ) for 6 people to do this task?*

- $$\frac{\frac{\text{Initial Workers}}{1\ \text{Task}}}{\text{Amount of Time Initally}} = \frac{\frac{\text{Workers in case 2}}{1\ \text{Task}}}{\text{Amount of Time needed}}$$

- $$\frac{\text{Wi}}{\frac{1}{\text{Ti}}} = \frac{\text{Wf}}{\frac{1}{\text{T2}}}$$

- $$\frac{4}{\frac{1}{7}} = \frac{6}{\frac{1}{\text{X}}}$$

- You could go through and first simplify it and then take the ratio of the numbers on one side once the variable is isolated, but the slide rule allows for this to be solved as is!?

- Take 4 on the D Scale and slide 7 on the C1 Scale over it. Start with the cursor here.
- Read along the C1 Scale to 6 and look below on the D Scale to find the answer 4.66 days.

- **In Physics, for example, say you have a balance beam.**

- *If on one side of the balance you have 26 g a distance of 32 cm from the center,*
- *how much mass must be placed on the opposite side at a distance of 20 cm from the center in the opposite direction so that it balances?*

- cw is clockwise, ccw is counter-clockwise

- $$\frac{\text{Mass cw}}{\text{Mass ccw}} = \frac{\text{Distance ccw}}{\text{Distance cw}}$$

- $$\frac{26\ g}{X\ g} = \frac{20\ cm}{32\ cm}$$

- X = 41.6 g

- This idea can be applied to problems in chemistry too.

- Take for example, conservation of mass in a problem where one has to determine the mass of a reactant product in a total mass size where one is only given a small sample for testing.

- $$\frac{\text{Mass of reactant in sample}}{\text{Mass of Sample}} = \frac{\text{Mass of reactant in total mass}}{\text{Total Mass}}$$

- This list applies to any and all sciences and is limited only by the imagination in the questions being asked.

## Using the CF & DF Scales ( $\pi$C & $\pi$D ) :

- The CF and DF Scales are called Folded Scales since they do not start at the Index, but instead are at a chosen point, namely here being $\pi$.
- Why $\pi$?

- Simple – there are many calculations that involve $\pi$ that extend from Circles, such as the circumference of a circle :

- 

- Take any value on the C or D scale and now look to the cursor value on the corresponding CF or DF scale. It is p times greater. This is the formula for the Circumference of a Circle : $C = \pi*d$, where d, the diameter of the circle is being read from the C or D scale and its circumference is then found on the CF or DF scale.
- For example – put the cursor of your slide rule on 3 on the D scale and find the DF scale. We are assuming we have a circle with a diameter of 3 units, so the question is : what is that circle's circumference? On the DF scale we read the answer of 9.42 units
- Note that the reverse is true as well. If we know the circumference of a circle, we can read it on the CF or DF scale and find its diameter quickly on the C or D scale.
- Another important role that a folded scale has is this : Since it starts at another point on the line, it is easier to do multiplication and division without having to change up Indexes as often.
- For example let's try 3 x 6
- We might first be inclined to slide the Left C Index over 3 on the D scale before realizing we should have first chosen the right C Index.
- But no bother –
- Read the value of 6 on the CF scale and find opposite it on the DF scale our intended answer, 18!
- What of the decimal point rules here, however?
- Going back to our original rules,
- First estimate the answer : Anything past 3.x will result in a value greater than 10, since by 3x4 we are at 12 already.
- Next, if we were to only use the C and D scales, then we would have had to use the Right Index instead of the Left one, hence we would have added one to our exponent total, or 1 in this case ( since the values started with exponents of 0 ).
- Another way to think of this problem is this : Since we have gone past the end of the scale we are on and moved on to the next one in sequence, it is 10x larger, so instead of being 1 to 10, it is now 10 to 100, so the value of 1.8 is really 18 in this case.
- Note : You can do multiplication and division with the CF and DF scales and the same rules apply for them as they did for the C & D scales presented previously. Realize that the index is still 1.

## Using the C1 Scale ( $\frac{1}{x}$ ) :

- This is the Inverse Scale. – See an example of use in combination problems – this is a common use of the C1 scale.
- It is basically C scale in reverse.
- It is written right to left ( regular are left to right )

- The inverse of any number can be found by aligning a cursor over a given value of C and on C1 is the inverse. ( Be sure to keep track of the decimal! )
- For example if reading '5' on the C scale, on the C1 it must be 0.2
- To multiply with the C1 Scale set one of the numbers on C1 over the other number on D.
- Read the Answer on the D scale under the Index on C1 scale.
- Any problem with a fractional component can be examined more easily with C1. − It can be a fraction or be the denominator of a given expression.
- Especially useful in combination problems. − see prior for example in the combination section under proportions.
- Use the rules for multiplication and division and read the proper index for a solution
- One of the other important and common uses of the C1 Scale is to be the scale when reading for a value from the tangent scale angles above 45° ( 45° to about 80° )

## Using the A & B Scales ( Squares $X^2$ & Square Roots $\sqrt{X}$ ) :

- The A & B Scales are double scales and are the squares of C & D Scales −
- That is to say it is double logarithmic (1 to 10 and 10 to 100)
- Conversely C & D are the square root values of A & B
- 
- Rules for Square Roots of Numbers :
- Note Left side is Left Index to Middle Index, and Right side is Middle Index to the Right Index
- 
- Odd Number Digit Rule :
- For Whole Numbers > 1
- If the number of digits left of the decimal point in the value being considered for square rooting is odd, Read the value from Left-Hand side of A
- For Numbers 0 < x < 1
- If the number of zeroes to the right of the decimal is odd then Read the value from the Left-Hand side of A
- 
- Even Number Digit Rule :
- For Whole Numbers > 1
- If the number of digits left of the decimal point in the value being considered for square rooting is even, Read the value from Right-Hand side of A
- For Numbers 0 < x < 1
- If the number of zeroes to the right of the decimal is even then Read the value from the Right-Hand side of A − Note this includes no zero at all ( just .X )

- To summarize the rule ( and make reading a R scale easier ) :
- The Left hand side of the A & B scales is for Odd Number of Digits or the Odd Number of Zeroes in a Number ( this corresponds to the R1 scale )
- The Right hand side of the A & B scales is for Even Number of Digits or the Even Number of Zeroes in a Number ( this corresponds to the R2 scale )

| For 0 < X < 1 | | | | | | | | | |
|---|---|---|---|---|---|---|---|---|---|
| No. of zeroes in between X and the decimal | 0 | 1 | 2 | 3 | 4 | 5 | 6 | 7 | Continue the pattern of even then odd values |
| Which Side ( L or R )? | R | L | R | L | R | L | R | L | Continue pattern R,L,etc |
| $\sqrt{\phantom{x}}$ Answer and the no. of zeroes between ans. And decimal point | 0 | 0 | 1 | 1 | 2 | 2 | 3 | 3 | Continue Pattern 4,4,5,5,etc |

| Number to have a square root taken | Answer on slide rule |
|---|---|
| 0.2 | 0.447 |
| 0.02 | 0.141 |
| 0.002 | 0.0447 |
| 0.0002 | 0.014 |

| For X > 1 | | | | | | | | | |
|---|---|---|---|---|---|---|---|---|---|
| No. of whole digits in value X | 1 | 2 | 3 | 4 | 5 | 6 | 7 | 8 | Continue the pattern of even then odd values |
| Which Side ( L or R )? | L | R | L | R | L | R | L | R | Continue pattern R,L,etc |
| No. of Digits in the answer $\sqrt{\phantom{x}}$ | 1 | 1 | 2 | 2 | 3 | 3 | 4 | 4 | Continue Pattern 5,5,etc |

| Number to have a square root taken | Answer on slide rule |
|---|---|
| 2 | 1.41 |
| 20 | 4.47 |
| 200 | 14.1 |
| 2,000 | 44.7 |
| 20,000 | 141. |

## Using the K Scale ( Cubes X³ & Cube Roots √X ) :

- The K Scale is a triple scale and is the cube of the values on C & D
- That is its range is 1 to 10, 10 to 100, and 100 to 1000.
- Rules for Taking Cubes :
- For the Rules consider the K scale divided into 3 sections from one index to the next.
- The sections are Left, Middle, and Right
- For All Values under consideration for cube root extraction :
- Divide the Number into groups of 3 ( starting at the decimal point ) and go left or right of the decimal point as needed in creating these groups
- Look at the Left-most group with non-zero digits in it
- If the group has 1 digit then use the Left portion of the K scale ( for example 1-10 )
- If the group has 2 digits then use the Middle portion of the K scale ( for example 10-100 )
- If the group has 3 digits then use the Right portion of the K scale ( for example 100-1000 )
- To continue further, continue counting in 3's.
- An easy way to find where a value falls if <1 :
- Let right-most 1 be 1 and go backwards as powers of 10 for any given decimal value :
- Example 0.00X is in the thousands place
- The right-most '1' is 1000/1000,
- The next '1' left of it is 100/1000,
- The next '1' left of it is 10/1000
- And the left-most is 1/1000
- If there are more restart at right-most and continue ( Note the scale wraps around then )
- The number of groups left of the decimal point determines where the <u>decimal point</u> in the <u>answer</u> falls :
- If there are two groups ( of 3 ) left of the decimal point ( complete or not ! ) ( means values 1,000 – 999,000 ) then there are 2 figures left of the decimal point in the answer.
- Values : X,XX0 to XXX,000
- If there is one group of 3 ( values 1-999 ) then the answer has one value left of the decimal point
- Values : X.XX to XXX.
- If there are 3 groups to the right of the decimal point ( where one or more falls in the tenths-hundredths-thousandths columns ) then the answer has 3 figures right of the decimal point starting just past the decimal point
- Values : 0.XXX to 0.00X XX-,---
- Values : 0.000 XXX to 0.000 00X XX0 ---
- If the number has four groups to the right of the decimal point where the first group is all zeroes and at least one value is in the ten-thousandths,

hundred-thousandths, or millionths place, then the answer will have 4 figures to the right of the decimal point where the first is a zero ( 0 ).

| Decimal value for X 0 < x < 1 | | Number of Zeroes in value | | |
|---|---|---|---|---|
| Which portion of K scale to read? | R, C, L | R ,C ,L | R ,C ,L | R ,C ,L |
| Q number of zeroes in value X | 0,1,2 | 3,4,5 | 6,7,8 | 9,10,11 |
| Number of zeroes in answer to $\sqrt{\phantom{x}}$ | 0 | 1 | 2 | 3 |

| whole values for X X > 1 | | Number of Digits in value | | |
|---|---|---|---|---|
| Which portion of K scale to read? | R, C, L | R, C, L | R, C, L | R, C ,L |
| Q number of digits in value X | 1,2,3 | 4,5,6 | 7,8,9 | 10,11,12 |
| Number of whole digits in answer to $\sqrt{\phantom{x}}$ | 1 | 2 | 3 | 4 |

# Using the L Scale ( Log $_{Base\ 10}$ (N) ) :

- The L Scale is the Log value of the C & D scales
- It is effectively used for determining powers and roots for a wide range of values. Note that the powers do not have to be whole numbers as well as roots can be any fractional value one is interested in determining
- The L Scale provides the log of values 1 to 10 with no changes
- If the Number is <1 ( includes <0 ) or >10, the log value has both the Characteristic ( the value before the decimal point ) and the Mantissa ( the value after the decimal point found on the scale )
- Note that the mantissa will be the same for a given value independent of the decimal point.

45

- 
- For example the log (2) mantissa is 301, as is for log (20), log (200), etc – the only difference is the Characteristic, so log(2) = 0.301, log(20)=1.301, log(200) = 2.301, et al
- When the number is <1, can use scientific notation to determine the value to add to the log of the number in question. Log(0.2) = Log(2 x $10^{-1}$ ) = 0.301-1 = -0.699
- 
- **Procedure below for logs in general :**
- First Look up the Log value for any given number treating it in Scientific Notation format so that it is >1 and <10
- Locate the characteristic of the scientific notation value on the C or D Scale and look below to the L Scale
- This L value is the Mantissa
- Add the Mantissa to the Exponent of the Scientific Notation exponent
- This Sum is the Final Answer
- The L Scale is useful for $X^N$ and $X^{1/N}$
- To solve – Look up X on C and find its log value on L
- Then Multiply or Divide as needed the N or 1/N value involved in the problem using the rules for multiplication or division. Be sure to watch the decimal point.
- This new value, Q, is computed
- Now Search for the Q Mantissa on the L scale
- Note that if there is a Characteristic, it becomes the power of 10 for decimal placement of the answer!
- Alternative to finding a log for values between 0 and 1
- If a desired value for log(X) has 0<X<1 as a decimal in the tenths place, the C1 scale can be used. Reading 2 on C1 the log value is the log of (0.2) for example.

## Using the S Scale ( sin(θ) & sin⁻¹(θ) ) :

- The Sine ( S ) Scale has values representing the angles from approximately $5.5°$ to $90°$
- When the cursor is on any value of the S scale ( the left number printed in black typically ) its sine value can be found under the cursor on the C or D Scale
- Note : Since Sine functions range from 0 to 1, all values read from C or D are decimal and begin as 0.XXX ( for most models )
- Note : Some Box-style Post Slide Rules have a cursor on the backside where the S scale often is and with a value on it the sine & tangent (T scale) of the angle is found on the B scale!
- For angles $0°$ to $5.5°$ there is often an ST scale and represents both Sine and Tangent functions as these have approximately the same values for small angles such as these ( also the decimal is 0.0XXX )

- 
- Many Sine Scales have a second set of numbers written in red to the right of the values on the scale –
- These are the complimentary angles hence they represent the Cosine of those angles listed in red
- Recall sin $(\Theta)$ = cos $(90°-\Theta)$
- To multiply by the Sine of a given angle simply place the angle over the appropriate D scale index
- Next read along D to the value to be multiplied by
- Find the Answer on the C Scale above that point on the D scale
- Remember to watch the decimal points since the C value is a decimal value
- To divide by the sine of an angle simply place the angle on S scale acting as the dividend over the divisor on the D scale
- Find the Answer at the C scale corresponding D scale Index
- If the square to the sine of an angle is needed since the sine value is on C, the square of the sine value is found on A.
- To find the log of an angle for a trigonometric function such as sine, look up the sine of the angle on C scale, then read on the corresponding L scale for the log of the sine of this angle.

## Using the T Scale ( tan($\theta$) & tan$^{-1}$($\theta$) ) :

- The Tangent ( T ) Scale has values printed in black from 5.5° to 45° when read left to right
- The values ( typically red ) read right to left are 45° to 84.5°
- Like the directions for Sine ( S ) Scale,
- To read the tangent $(\Theta)$
- place the cursor on the desired angle
- If the angle is from 5.5° to 45°, the Tangent to this angle is found on the C or D scale and is read as a decimal 0.XXX
- Recall that tan(45°) = 1.00
- In fact this value lines up with the Right Index of C & D
- If the angle is 45° to 84.5° read the T Scale from Right to left to find the value and place the cursor there
- Answer to the Tangent of this angle is found on the C1 Scale
- The values on the C1 scale are read as whole numbers as the line appears
- ( Note there are slide rules that have two T lines, hence are read from left to right on the C scale and one must keep in mind that the C values start at 1 ( tan 45° )
- Like the S scale since the tangent ( like the sine value ) value is on the C scale the rules for multiplying and dividing are the same as noted in the S Scale section
- Note since tan$(\Theta)$ = 1/cot$(\Theta)$ one can compute these as needed

## Using the ST Scales :

- If your Slide Rule has a ST Scale it can be used for small angles for both the sine and tangent function as these are very close to each other in value between 0.6° and 5.5°
- Much like S Scale and T Scale readings the angle is on the ST scale and the reading comes from the C Scale.
- Important Reading Scale Note : Here however the readings range from 0.01 and 0.1, so the reading has $10^{-2}$ as its exponent.
- What if there is no ST Scale :
- In approximations :
- We can use : $\sin(x) = \tan(x) = \dfrac{x}{\frac{180}{\pi}} = \dfrac{x}{57.3}$
- Note that this approximation also is the radian value for that angle!
- Here x is the small angle and using the C and D scales then provides a reasonable value for the sine or tangent of the small angle ( within the same range ).


## Using the Log ( LLn Scales ) :

- The LLn Scales are used to raise a number in question to a power or find a root of that number.
- Each line is a power of e ( 2.718... ), some models it is the power of 10, of the next line in the list ( LL1, LL2, et al ).
- That is to say each line is e to some power ( from -10X ( LLO3 ) to +10X ( LL3 ) )
- The Table of Names and Powers for LL scales are below showing both the power and the range of values to be found on those scales. :
- Looking at the tables, it is clear that the spacing needs to be examined carefully when reading the scales. Be sure to look at the two primary values that your values is between ( as an example ) and then look at the number of secondary divisions to determine the appropriate value.
- These scales are indispensible when looking for any given power or root as needed ( not necessarily whole number ones either ).
- Also since each is 10x the line before it, for example LL3 is 10x LL2, and LL2 is 10x LL1, and so on, hence LL3 is 100x LL1. This means that for any power of 'e' which is used for the scale when reading from the C or D scale each is 10x the others.
- For example $e^2$ is found by looking at the D line for 2 and reading from the $e^{1-10x}$ line, which is LL3 and is approximately 7.4
- Better still, what if you wanted the inverse of $e^2$ or 7.4? Simple read its value on LL03 – Be careful in reading the scales, these are in reverse first off and have decimal values – so read it from right to left
- Here the inverse of 7.4 is between 0.1 and 0.2 and reads 0.135 – try this yourself and practice reading the scale properly.

- 
- What about $e^{0.2}$, that is found on the LL2 scale, ( yields a value nearly 1.222 )
- and $e^{0.02}$ is on the same cursor line with the cursor the whole time on 2 on D.
- This is one of the great values of the LLn scales. Any value from 1.001 to nearly 100,000 can be used from LL0 to LL3 and its inverse is present as well if needed. That would be $10^{-5}$ to 0.999 found on LL00 to LL03. It is typically written in this range though, due to the effectiveness of the slide rule : having a range of values ( up to 80 inches in length on these LLN scales ) of running from 0.00005 to 20,000.
- Looking at a value on say LL0 scale and reading the value below on LL1 scale it becomes that number raised to the power of e 10x more. So an increase in number is 10x the prior line.
- The LL0 value if read with the LL2 scale will then be that value raised to the power of 100.
- Going from a higher LL scale number to a lower one means that it is $1/10^{th}$ power per each jump.
- To raise to a negative ( $-$ ) exponential power : This can be done directly or find the positive value then its inverse on the corresponding negative exponent line.
- If raising negative exponent to positive exponent values :
- If one wants to find 10x the value in power simply read from one line to the next, such as going from LL1 to LL2 in essence multiplies the power by 10 if needed.
- Going in the reverse direction divides the power by 10, of course.
- For example read a value on $-$LL1 scale and find that value to the $10^{th}$ power on $-$LL2 scale.
- 
- Note : the $-$LL Scales are normally written in red. And the values increase from right to left ( as do all other inverse scales )
- Recognize that $-$LLX scales are the inverse of LLX scales correspondingly. So the inverse of the values found on LL1 are found on LL01 also known as $-$LL1.
- 
- **<u>Using the LLn Scales in general :</u>**
- 
- To find $X^N$ :
- Look up the value X on the LL Scale.
- Slide the Index of the C ( 1 ) scale over it.
- Move the cursor along the C Scale to the power ( N ) and
- Read below the answer on the needed LLX scale
- For example, 2 is squared, 3 is cubed, 5 is the $5^{th}$ power – but you can do any power in between too – such as $4.7^{th}$ power for whatever reason is needed :
- ( be sure to estimate since it may have gone from one of the LLX Scales to the next one in line )

- 
- **<u>One could find roots as well :</u>**
- For a given value, X find this on the needed LLX scale
- Place the needed root over it ( N ) on the C scale
- Note : Recognizing the this is read as 1/N mentally
- Now read in conjunction with the C scale Index the value on the appropriate LLX scale the root value
- Essentially taking the root is the reverse of the process of determining a power – much like doing multiplication and division with the slide rule in terms of directions.
- Note : this process works for –LL Scales as well.
- LL Scales are not good for numbers very close to 1, such as 1.001 or 0.999.
- There is an approximation for this value for small values of 'n' with :

$$( 1 + n )^p = 1 + d^p$$

| Name | Power | Range | Name | Power | Range |
|------|-------|-------|------|-------|-------|
| (+)LL3 | $e^{+1.0x}$ | e to 22k | (-)LL03 | $e^{-1.0x}$ | 1/e to 1/22k |
| (+)LL2 | $e^{+0.1x}$ | 1.1 to e | (-)LL02 | $e^{-0.1x}$ | 0.91 to 1/e |
| (+)LL1 | $e^{+0.001x}$ | 1.01 to 1.1 | (-)LL01 | $e^{-0.001x}$ | 0.990 to 0.91 |
| (+)LL0 | $e^{+0.0001x}$ | 1.001 to 1.01 | (-)LL00 | $e^{-0.0001x}$ | 0.999 to 0.990 |

Note : in table, the + and – denotation is used together, such as –LLN and +LLN, on some slide rules, while on others it is LLN and LL0N denotation depicted

## Tricks & Tips in Slide Rule Use :

- Check alignment of scales and adjust as needed
- Always be sure of the level of precision of the scales for proper reading. Most 10" Slide Rules are quite good to 2 significant digits with even a 3$^{rd}$ estimated significant digit being possible.
- In terms of all calculations undertaken, Be sure to first estimate the answers ahead of time mentally
- Often it is a good idea to convert most values to scientific notion both for ease of calculation and determination of decimal placement.
- 
- **Summary of Scientific Notation Rules :**
- **The Rules for Decimal Placement :**
- 1) Always first estimate the Answer.
- 2) Convert all values into Scientific Notation.
- 3) For Any Multiplied Values, Add up the Exponents and For Any Divided Values, Subtract the Exponents.
- 4) If for any single Multiplication operation the Slide moves to the Left, then add +1 to the exponent total ( Why? Because in essence, we have gone off this scale and are adding two values that extend beyond the scale in front of us to a next one in line, which is 10x the line before it )
- 5) If for any single Division operation and the Slide moves to the Right, then -1 from the exponent total ( Why? Because in essence, we have gone off this scale and are subtracting two values that place the answer on the scale to the left of the one in front of us, which is 1/10th the line before it )
- 6) Mentally treat any two values independently ( just as whole numbers now ) but keep track of what the exponent will be through the rules.
- 7) That is to say answer the question as to where the Slide has moved and factor in that addition or subtraction as needed for each operation independently.
- 8) Take this final answer and use the exponent figure arrived at to determine the decimal placement!

- **Alternative Decimal Place Method :**
- **Summary of Decimal Rules using the Counting Digits Method found in some slide rule books :**
- The key to the digits method is to tally the number of digits in a number.
- The basic rule for this is this :
- For numbers X > 1 the number of digits is simply the number of digits in the number.
- For example : 7 has 1 digit, 70 has 2 digits, and 70,000 has 5 digits, etc.
- What if the number is greater than 1 but has decimals, though?
- For example the number is 23.45, how many digits does it have?
- It has only 2.
- In the case of X > 1, all decimal values are overlooked in the digit count.
- So the next question then is, what of values 0 < X < 1 then, what is their digit count?

- 
- In the Digit Count Method all values 0.1 to 0.9 expressed as nothing more than 1/10ths values have 0 ( zero ) digits.
- So 0.6 has 0 digits, for example.
- With each decimal place it is like a reverse number line :
- 0.01 has -1 number of digits,
- 0.001 has -2 number of digits,
- 0.0001 has -3 digits,
- And so on...
- To help remember this the value of digits is essentially the negative number of zeroes past the decimal point.
- Summary of these ideas on Counting :
- For numbers greater than 1, count all the numbers up to the decimal and treat these values as positive. For numbers less than 1, count the number of places up to the first nonzero number and view these as negative values. Sum up these numbers in multiplication and when the slide extends to the right, subtract one from the sum.
- 
- With the number of digits in all the values you have in your problem now look to the following rules when it comes to multiplication and division :
- 
- Multiplication with C & D Scales :
- 1) If the slide projects to the right of the stock during multiplication, the digit count for the product is one less than the sum of the digit counts for both the values in your calculation ( called the multiplicand and the multiplier ).
- 2) If the slide projects to the left of the stock, the digit count for the product is equal to the sum of the digit counts for the multiplicand and the multiplier.
- Division with the C & D Scales :
- 1) If the slide projects to the right of the slide rule stock during a division, the digit count for the quotient is one more than the digit count for the dividend ( the numerator ) minus the digit count for the divisor ( the denominator ).
- 2) If the slide projects to the left of the stock, the digit count for the quotient is equal to the digit count for the dividend minus the digit count for the divisor.
- Multiplication with the C1 & D Scales :
- 1) If the slide projects to the right of the stock during multiplication, the digit count for the product is equal to the sum of the digit counts for the multiplicand and the multiplier.
- 2) If the slide projects to the left of the stock, the digit count for the product is one less than the sum of the digit counts for the multiplicand and multiplier.
- Division with the C1 & D Scales :
- 1) If the slide projects to the right of the stock during division, the digit count for the quotient is equal to the digit count for the dividend minus the digit count for the divisor.
- 2) If the slide projects to the left of the stock during division, the digit count for the quotient is one more than the digit count for the dividend minus the digit count for the divisor.

- 
- <u>Other Ideas :</u>
- 
- Examine the problem carefully, working from the inside out, and use the best scale for that calculation
- Recognize that values can be found easily with given scales – squaring, square rooting, cubing, cube roots, sine values, etc. When moving the cursor its alignment may find it at a place to make use of these as needed.
- Proportions ( which also conversions and 3-variable functions can be treated as ) are very straightforward and easy to set up and solve on a slide rule.
- Note that most things are Ratios or Proportions and the Slide Rule is the best tool for these calculations since it sets up all similar ratios instantly with any setting!
- Create a needed scale! Take the activity where we used the C & D scales with the C1 scale. If there is no C1 scale, simply invert the slide in the slide rule and use the reversed C scale as if it were a C1 scale now.

- **<u>How to Add Numbers on a Slide Rule :</u>**
- Suppose we have two numbers X & Y so that we want their sum ( X + Y ) and yet use a slide rule!
- Let's first rearrange this expression : $X*( 1 + \frac{Y}{X} )$
- With a slide rule simply set the Y value on C over the X value on D.
- 
- If Y > X read answer above the D index from the C scale as whole number with decimal and add one to it. Note that the decimal placement rules also must apply however.
- OR
- If Y< X read answer at D index on the C scale as decimal value and add one so that it is 1.***. Note that the decimal placement rules must also apply however.
- 
- Now Take the result and find it on D and place this value under the left C index.
- Now, Regardless of Y > X or Y < X, now read along the C scale to the X value and find the answer for the sum of X + Y on the D scale below! Be certain to have an estimate of your answer and employ the proper decimal placement rules for reading the slide rule.

- **Other Things to find on the Slide Rule :**

- There are many ratios that can give good estimates in situations :

- Set up these ratios and find the known value and opposite it is the sought after answer -

$$\frac{\text{Circumference of Circle}}{\text{Diameter of Circle}} = \frac{355}{113} \qquad \frac{\text{Feet}}{\text{Meters}} = \frac{82}{25} \qquad \frac{\text{Atmospheres}}{\text{Feet of Water}} = \frac{23}{780}$$

$$\frac{\text{U.S Gallon}}{\text{Cubic inches}} = \frac{1}{231} \qquad \frac{\text{Feet of Water}}{\text{Pounds per sq inch}} = \frac{60}{26} \qquad \frac{\text{Side of a Square}}{\text{Diagonal of a Square}} = \frac{70}{99}$$

$$\frac{\text{US gallons}}{\text{Liters}} = \frac{14}{53} \qquad \frac{\text{US Gallons}}{\text{Imperial gallons}} = \frac{6}{5} \qquad \frac{\text{Inches of Mercury}}{\text{Feet of Water}} = \frac{15}{17}$$

$$\frac{\text{Yards per Minute}}{\text{Miles per Hour}} = \frac{88}{3} \qquad \frac{\text{Pounds per sq yard}}{\text{Kgs per sq meter}} = \frac{46}{25} \qquad \frac{\text{Weight of fresh water}}{\text{Weight of sea water}} = \frac{38}{39}$$

$$\frac{\text{US Gallons of Water}}{\text{Weight in pounds}} = \frac{3}{25} \qquad \frac{\text{Ounces}}{\text{Grams}} = \frac{6}{170} \qquad \frac{\text{Diameter of Circle}}{\text{Side of equal square}} = \frac{79}{70}$$

$$\frac{\text{Inches}}{\text{Centimeters}} = \frac{26}{66} \qquad \frac{\text{Pounds}}{\text{Kilograms}} = \frac{75}{34} \qquad \frac{\text{Area of Circle}}{\text{Area of Inscribed Square}} = \frac{322}{205}$$

# Ch.V
# Gauge Marks and Scales of the Slide Rule

## The Scale

The primary use of the slide rule is seen in our activities for multiplication and division. This was done by using non-linearly divided scales that are divided instead in a logarithmic manner ( that is to say the numbers were not evenly spaced like on a ruler but here by logarithmic distances ).

Why use the logarithmic spacing of the numbers ? The most basic linear form utilizes two scales where the numbers are logarithmically spaced. Due to the mathematical properties of logarithms, the spacing of the numbers on these lines allows for easy and rapid multiplication and division with the same set of rules despite the type of slide rule used. ( This idea is explored in the history section in more detail discussing the discovery and importance of logarithms to history which the slide rule can be seen as the physical visual manifestation of this math form ).

Many slide rules have more than 2 scales, and these can be used for many other mathematical operations as we have seen, such as : squaring, square roots, cubing cube roots, raise to various powers, taking a various root of a number, common and natural logs, and trigonometric values such as sine, cosine, and tangent.

Standard scales spaced in a logarithmic fashion, like C and D are said to be **Single Scales**. That is, there is one run from 1 to 10 in the distance for the slide rule, such as 25 cm ( i.e. 10 in ). If there are two scales running from 1 to 10 in the same distance, it is said to be a **Double Scale**, such as for A or B scales. Of course if there are 3 runs of 1 to 10, which is the case for the K scale, this is a **Triple Scale**.

Another type of scale is the **Folded Scale**. Instead of starting at the Index point of 1, it begins at or near another point on the logarithmic line. The usual choice is at pi ($\pi$). The choice of pi was two-fold. One it allowed for using pi as an Index, so that calculations involving circles and cylinders could easily be done and two, by cutting the original C or D scale at a different point, made alignment more convenient for times when multiplication involved numbers from each end of the regular fundamental scale.

The C and D scale are the primary or fundamental scales of the Slide Rule and all others are based on them. For example, we see from the activities, the tangent ( as well as the sine ) scale is read in relation to the C or D scale. This idea is further illustrated in the table which lists in alphabetical order the scales, their mathematical relationship to C or D and a brief description of that scale.

## Table of Scales and Uses

| SCALE | Mathematical Relation to C or D | DESCRIPTION and Other Notes |
|---|---|---|
| **A** | $x^2$ | Square Values of Fundamental D scale<br>Double scale, Opposite to B scale |
| **A1** | $1/x^2$ | Reciprocal of A scale,<br>Reciprocal of square of D scale |
| **B** | $x^2$ | Square Values of Fundamental D scale<br>Double scale, Opposite to A scale |
| **C** | $x$ | Fundamental ( Single ) Scale<br>On Slide opposite to D scale |
| **CF** | $\pi x$ | Folded Fundamental Scale, starts at $\pi$<br>Opposite to **DF scale** |
| **CI** | $1/x$ | Reciprocal of Fundamental Scale<br>On Slide – Basically a Reverse Scale |
| **CIF** | $1/\pi x$ | Reciprocal of CF scale<br>Reciprocal of Folded Fundamental Scale C |
| **D** | $x$ | Fundamental (Single ) Scale<br>On the Stock opposite to C scale |
| **E** | $e^x$ | Log-log scale – see LL3 |
| **K** | $x^3$ | Cube Values of the Fundamental D scale<br>Triple Scale |
| **L** | $\text{Log}_{10}x$ | Mantissa of the common logarithm of Fundamental D scale value |
| **LL** | $\text{Ln}(x)$ | Mantissa of the natural logarithm of the Fundamental D scale value |
| **LL0** | $e^{0.001x}$ | Scale yields 'e' raised to 0.0001*x power, where x is read from the fundamental scale.<br>Positive Log-log scales are used to raise a number to some exponent or find roots >1<br>LL scales are to enable one to raise values to a power and take roots very readily |
| **LL1** | $e^{0.01x}$ | Scale yields 'e' raised to 0.01*x power, where x is read from the fundamental scale. |
| **LL2** | $e^{0.1x}$ | Scale yields 'e' raised to 0.1*x power, where x is read from the fundamental scale. |
| **LL3** | $e^x$ | Scale yields 'e' raised to x power, where x is read from the fundamental scale. |
| **LL00** | $e^{-0.001x}$ | Scale yields 'e' raised to (-0.001*x) power, where x is read from the fundamental scale.<br>Negative power means $1/e^{-0.001x}$<br>Negative exponent log-log scales are used to raise numbers to a power or find roots <1 |

| | | |
|---|---|---|
| **LL01** | $e^{-0.01x}$ | Scale yields 'e' raised to (-0.01*x) power, where x is read from the fundamental scale. Negative power means $1/e^{-0.01x}$ |
| **LL02** | $e^{-0.1x}$ | Scale yields 'e' raised to (-0.1*x) power, where x is read from the fundamental scale. Negative power means $1/e^{-0.1x}$ |
| **LL03** | $e^{-x}$ | Scale yields 'e' raised to (-x) power, where x is read from the fundamental scale. Negative power means $1/e^{-x}$ |
| **P** | $(1-(0.1x)^2)^{1/2}$ | Pythagorean Scale. Cosine of $\sin^{-1}$ D scale |
| **R1** | $\sqrt{\ }$ | Square root of Scale D value. Scale twice the length of D scale. R1 runs 1 to 3.2, R2 runs 3 to 10. Aka: W1, W2 and Sq1, Sq2 |
| **S** | $\sin^{-1}x$ | **Scale D** value is the sine of the angle in degrees read on the **S scale**. Runs 5.7° to 90° |
| **ST** | Sinx, Tanx | Sine & Tangent of small angles 0.58° to 5.73°. Same and S&T scale. Used since sine and tangent are similar at these angles |
| **T** | $\tan^{-1}x$ | **Scale D or C and CI** value is the tangent of the angle in degrees read on the **T scale**. Runs 5.7° to 84.3°. D scale read for angles 5.7° to 45° increasing. CI scale read for angles 45° to 84.3° ( printed in red and in reverse order ) |

The table above is by no means all of the scales, nor does it trace the differences in names for the same scales as they changed through time. It is a representation of the most common scales and represents a list that most slide rules commonly have a subset of.

As noted earlier, one could simply use a 9 scale form ( such as in the activities ) and be able to compute the overwhelming majority of problem types that even a regular scientific calculator is capable of today.

This does not mean that the other scales were merely for show, though there was probably some pride and showing off in the office if one had a more advanced model I imagine. The other scales were applicable to various disciplines, such as chemistry, physics, electrical engineering, and the like.

This idea is particularly true when one considers not just the right scale combination form, but also relevant and useful gauge marks that may be on the scales. Some are listed and described below in the accompanying table. Like the scale table, this is not complete and represents a cross-sectional view of them.

## Table of Gauge Marks on the Slide Rule

Besides the numbers on a slide rule perhaps you have noticed some out of place marks or letters? These are gauge marks. These marks are at specific places corresponding with their value and have a given prescribed mathematical value to the user of the slide rule.

The table below lists some of the more common ones. This table is very far from complete but illustrates some of the possibilities. Note that not all makers use the same letters nor do they have all of the same marks on their various models. Some were probably put on there since those models may have been marketed to a particular set of professions where that mark would have value and use.

Look at some of the examples such as symbols for the weight of copper conductors, watts in one horsepower which clearly had applications in the electrical industry. Other constants as gauge marks can include the acceleration due to gravity for calculations in physics.

More common ones include the conversions of radians to degrees and vice versa as angular measures in calculations were commonplace.

Probably as no surprise the most common gauge point is $\pi$, since it can be used for all circular calculations. Using $\pi$ we realize the value in this. Knowing where it is at allows quick calculations of either multiplication or division as needed.

| Gauge Marks | Meaning | Value |
|---|---|---|
| C | Square root of $4/\pi$ Circle area calculations | 1.128 |
| G | Acceleration due to Gravity ( metric ) | 9.8 m/s$^2$ |
| G | Acceleration due to Gravity ( English ) | 32.2 ft/s$^2$ |
| L | Natural log of 10 Convert $\log_{10}$ & $\log_e$ | 2.3026 |
| R, p, or r | $180/\pi$ Convert radians & Degrees | 57.3 |
| p$'$ | Minutes in a radian | 3438 |
| Q | Radians in one Degree | 0.01745 |
| | $\pi/4$ | 0.7854 |
| $\pi$ | Pi – ratio of circle Circumference to Diameter | 3.1416 |
| W | Weight of copper | 111000 |
| - | Watts in one Horsepower (hp) | 745.47 (746) |

There are also common conversion marks and value marks with no special symbols on various slide rules, such as on the Fowler's Circular Calculators. Many of them have value marks for square-root of 2, square-root of 3, along with pi, and conversions for inches to centimeters, kilograms to pounds, square centimeters to square inches, and the like.

# Activity #1
## Using Math ( and the Slide Rule ) in Everyday Life
Grade Level : Middle School
Math Level : Calculating

## Everyday Life Calculations with the Slide Rule Activity

This Activity is the mathematical exploration of everyday life – such as determining miles per gallon, cost per unit ounce and the like, but all using a slide rule. Even if you don't, though I recommend it, these are basic things we should all know and engage in mentally to some degree. It is good to practice common math sense! : )

- o All of these calculations require either a question on the part of the person to consider the hypothetical or the actual items at hand to work on for calculations.
- o The materials needed depend on what is being done.
- o Know the basic rules for multiplication and division on a slide rule ( they are summarized below and if needed in the first question – miles per gallon ).
- o When scales other than C & D are used the rules are explained.
- o In most cases, only the C scale and D scales are needed.
- o Always estimate an answer and watch decimal placement.
- o Not always needed but is handy is a small pad of paper.

## Slide Rule Basics

1) For almost all calculations in this Activity, the C & D scales are used.
2) If considering a Ratio or Fraction, It is best to see the C Scale as the Numerator since it sits atop the D Scale and the D Scale as the Denominator
3) To Multiply, Place the Right or Left Index as needed of a given scale, say the D Scale below the first number to be multiplied on the C Scale.
4) Next, read along the D scale to the other number to be multiplied by and read above it the answer on the C scale!
5) If the number is not available, use the other index ( Right or Left ) and start the process over as needed.
6) If Dividing simply place the Numerator value on the C Scale over the Denominator Value on the D scale.
7) Read the Answer opposite the Index of the D Scale on the C Scale.
8) What about the use of Scientific Notation?
9) It is best to consider all values in Scientific Notation in all calculations. This is done by setting the decimal point past the first non-zero number in the value under consideration and multiplying by 10 raised to the appropriate power.
10) The power is determined to be positive if the decimal needs to move to the right to obtain the original number ( 2000 is $2 \times 10^3$ for example ), and negative if the decimal needs to move to the left to achieve the original number ( 0.002 is $2 \times 10^{-3}$ for example ).
11) When multiplying with values in Scientific Notation, simply add the exponents ( Result = Operand 1 Exponent + Operand 2 Exponent ).

12)

13) When dividing by Values in Scientific Notation, simply subtract the Denominator Exponent from the Numerator Exponent ( Result = Numerator Exponent – Denominator Exponent ).

14) Note the special case rules for linear slide rules :

15) If Dividing and the Slide Projects Right, Then Subtract One from Your Exponent Total to achieve the correct decimal placement.

16) If Multiplying and the Slide Projects Left, Then Add One to Your Exponent total to achieve the correct decimal placement.

17) Here is a full summary of decimal placement :

   ➢ **The Rules for Decimal Placement :**
   ➢ 1) Always first estimate the Answer.
   ➢ 2) Convert all values into Scientific Notation.
   ➢ 3) For Any Multiplied Values, Add up the Exponents and For Any Divided Values, Subtract the Exponents.
   ➢ 4) If for any single Multiplication operation the Slide moves to the Left, then add +1 to the exponent total
   ➢ 5) If for any single Division operation and the Slide moves to the Right, then -1 from the exponent total
   ➢ 6) Mentally treat any two values independently ( just as whole numbers now ) but keep track of what the exponent will be through the rules.
   ➢ 7) That is to say answer the question as to where the Slide has moved and factor in that addition or subtraction as needed for each operation independently.
   ➢ 8) Take this final answer and use the exponent figure arrived at to determine the decimal placement!

## Miles per Gallon Measured

1) For this calculation it is best to begin with a full tank of gas. Once full write down the number of miles on the Odometer [called the starting miles] ( or reset the trip odometer to zero ).

2) Drive for some period of time ( it can be a day, but if you have a regular work schedule it may be better to go for 2 or 3 days ).

3) Needless to say you put more gas in some time later. At that time write down 2 numbers.

4) First the current reading of the Odometer now [called the ending miles] ( or look at the trip odometer and write down the number of miles driven ).

5) Second write down the exact amount of gas put in this time. This is the number of gallons used.

6) The 10" slide rule will give reasonable results despite the number of digits in your numbers, but it is best to round off as needed to facilitate the calculation.

7) If you did not use the trip odometer, determine the difference between your starting miles and your ending miles by subtracting them. This is your trip total.

8) Quick review for Division and Multiplication on a Slide Rule :

9) To Divide place the numerator read from the C scale over the  denominator on the D scale and read the answer opposite the D scale index on the C scale.

10)

11) To Multiply place the left ( right if needed ) Index of the C scale over the first operand read from the D scale. Read along the C scale to the other operand and read opposite it on the D scale to find the Answer.

12) To determine the <u>average miles per gallon</u> :

$$\text{Mpg} = \frac{\text{trip total}}{\text{number of gallons used}}$$

13) This value is important in many ways – On long trips and the car is gassed up, you can now estimate the range the car will go before filling up by *multiplying the Mpg by the number of gallons*.

14) On a map, knowing the distance to some place an estimate of the amount of gas needed can be determined by *dividing the Trip Miles by the Mpg*.

15) Also estimated costs can be found by taking *the number of gallons used and multiplying by the average cost of gas at that time*.

16) This calculation can go along with the Average Speed one as well.

## Average Speed along with Distance and Time Determinations

1) To find the average speed what is needed is the total distance travelled and the total time taken for the trip.

2) This calculation can be done for cars, bikes, walking, boats, or any moving object under consideration.

3) If by car, either determine the total distance to be traveled by examining a map ( paper for those who like to figure it out themselves or on the internet and a map system like Google Maps ) OR

4) Actually travel the distance and either reset the trip odometer before the trip or write down the current odometer reading ( starting miles ) and then at the end of the trip write down the new odometer reading   ( ending miles ).

5) In either case of the trip, it must be traveled and times.

6) For the traveled trip be sure to not only keep track of miles but use a watch or a timer. On a watch mark down the start and end times.

7) Determine the miles traveled by subtracting the starting miles from the ending miles ( Miles Traveled ) and

8) Do the same for the amount of time to travel by subtracting the start time from the end time ( Travel Time ).

9) The Average Speed is found by :

$$v = \frac{\text{Miles Traveled}}{\text{Travel Time}}$$

10) Note that one does not need the total miles and total times to determine an average speed for part of a trip.

11) If you are on a long trip and had the start time and start miles down, then at any point in the trip the average speed for that portion of the trip can be determined.

12)

13) This might be useful in the case of a very long drive where if the total miles to cover is known and the average speed is determined for some point in the trip, then the amount of time needed is found by :

14) First subtract the miles covered from the total miles. *Take the remainder and divide this value by the average speed. This will give the amount of time in hours left for the journey.*

15) Estimated times can be determined for an entire trip by first estimating the average speed and dividing it into the total miles of the trip. This gives the number of estimated hours for the trip.

16) If there is a situation where the average speed is known and the amount of time is also known then *the distance traveled is simply the average speed times the amount of time traveled.*

17) Always watch to see that units match up as needed! If not convert units to agree with the problem at hand.

18) A good exercise is to convert units for those who like math practice. For example, convert mph to miles per second or feet per second.

**19)     Miles per Second = mph / 3600**

**20)     Feet per Second = Miles per Second * 5280**

21) Also why not try the metric conversions as well, such as mph to km/hr and then convert these to m/s.

**22)     Kilometers per Hour (km/hr)= 1.6(09) * Miles per Hour**

**23)     Meters per Second = (Km/hr) / 3.6**

24) These ideas apply to taking a flight too. Know the distance flown and time the trip to determine the average speed of the plane.

25) Note that distances do not have to be miles, it can be any measurement unit. A bicycle can be looked at from the number of houses it passes in a given time, for example. A crawling insect might be done in centimeters/minute.

## Cost per Unit Known ( Mass, Volume, Weight, Item Number )

1. Most stores today will provide the cost of an item per unit ounce, per pound, per unit number of the item. The first question then to determine is this : Is it correct?
2. The next goal of this activity is to compare it to other brands or the same brand on the shelf only packaged in a different size.
3. Note that often the larger sizes are not labeled in the same manner of cost per unit as the smaller ones.
4. Be sure to read the labels carefully since the unit cost provided might be something like cents per ounce yet the material is labeled in dollars and pounds.
5. Therefore it is a good idea to keep in the back of one's mind basic conversions, such as :
   - **1 pound = 16 ounces**
   - **1 cup = 8 ounces**
   - **1 gallon = 4 quarts**
   - **1 quart = 2 pints**
   - **1 pint = 16 ounces**
   - **1 kilogram = 1000 grams**
   - **1 kilogram = 2.2 lbs ( pounds )**
   - **1 ounce = 28.3 grams**

6. With these in mind it is best to choose base units for calculation and comparison, for example turn all units to ounces for mass or volume considerations.
7. Reading a label first determine the total number of ounces if not given.
8. To calculate the cost per unit ounce (CPO):

9. **Cost per Unit Ounce (CPO) =** $\dfrac{\text{Cost (\$)}}{\text{No. of Ounces}}$

10. For all other types of costs per unit whatever it is simply the cost divided by the number of ounces ( number of pounds ) (number of items ) and the like.
11. In all cases these are rates and a rate is a ratio or fraction.
12. The best way to treat all ratios is to let the Numerator ( here Cost ) be the C scale and the Numerator be the D scale ( here the number of Ounces ). The answer is read opposite the D scale index on the C scale.
13. Once you know the cost per item it is easy to determine the total cost of a purchase by taking that ratio and multiplying by the number of items you intend to buy.

## Computing Sales Tax or Determining the Tip

1) This activity is not a mistake though these two things are different circumstances. It turns out they are calculated in the same way on the slide rule.
2) For both you are mathematically taking a cost and adding to it a percentage of the cost.
3) Read the Cost on the C scale and place the left index of the D scale beneath it.
4) Now read along the D scale to the known sales tax rate or the desired amount of tip.

5)
6) Here each mark on the slide rule represents 1% or 1/100.
7) For example, The second mark is 2%, the fifth mark is 5%.
8) The easy mistake is misreading the D scale. Be sure to keep track of the decimal point here.
9) For example if the cost were $20 the first mark past the left index on the D scale has a reading of $20.20, the fifth mark is $21.00
10) Yet if the cost had been $200, then the first mark has a reading of $202 and the fifth mark has a reading of $210.
11) This illustrates the power of the slide rule since all numbers can be represented on this logarithmically-spaced number line and it is merely the decimal point that needs to be tracked.
12) When you read to 1.1 this is 10%. The value above it on the C scale is 10% greater than the original value.
13) If you want the amount of tax or tip you have to subtract the original cost from this new value to see how much the tip is.

## Calculating Cost of a Sale Item ( given percent off )

1) Many of us have been to stores proclaiming some given percentage off and large tables on cards describing what one pays.
2) With a slide rule this is a very easy task.
3) First, we need to acknowledge the trick in advertising. If it is 10% off, we are still paying 90% ( 100%-10% ), and we pay 75% if it is 25% off ( 100%-25% ) and so on.
4) Knowing this makes the task of finding our actual cost very easy and we can even find our savings too.
5) Read the Original Price on the C scale and place it over the Right Index of the D scale.
6) Now read from right to left along the C scale and find the percentage you are paying.
7) For example, if you have 10% off, read to 9 which is now considered 90% and read above it on the C scale your price.
8) If you have 25% off read to 7.5 since this is 75% and read the price above on the C scale.
9) What if you want the amount of savings? Instead of the right index, place the Price over the Left Index of the C scale.
10) Reading from left to right find the percentage off and find its value on the D scale above.
11) Note that you may have to use the Right or Left Index differently if the values go off the scale!
12) Notice the speed of this reading. In a single moment one can determine what any percentage is for any given value. Try that one with a calculator for those who think that faster!
13) By the way go back to the sales tax activity above if there is tax on it to determine the total cost if needed.

## Computing Pay Amount from Rates of Pay

1) For estimated gross pay, take the rate of pay ( R ) on the C scale
2) Place this value over the index on the D scale
3) Read along the D to find the Number of Hours worked ( H )
4) Find the answer on the C scale above ( G )
5) Notes : First note that the pay rate has to be in the same time unit as is multiplied. Second, this is the gross without any withholding. To factor that in take an estimated percentage of pay withheld for taxes and use the percent off calculation to find an estimated Net Pay. Third, if there is overtime pay, be sure to calculate those hours at that pay rate and add that to the base gross pay.

### G = R*H

## Estimating the Electric ( Water ) Bill from Meter Readings

1) This activity here examines directly the process of reading the meter, taking the values there and finding an estimated bill (excluding taxes and other fees on the bill).
2) Depending on the meter you have for the Electric and Water Meters, this will affect how it is read.
3) There are some that give a direct reading of the current level of consumption, and if present then you can proceed with the calculation.
4) If the number is not directly on it, then you need to start on the first day of the month if you want to monitor it for the month. ( If for a week, pick any day ).
5) Read the dial carefully. Most have dials and these go in opposite directions with each dial, first clockwise, then counterclockwise, etc.
6) When the arrow is between numbers be clear on how to determine what value it is. The arrow is always the value it is coming from and until it reaches the next value on the dial.
7) If the meter is in a dial pattern, be sure to start with the largest place value and work your way around the dial for the other digits.
8) In both the Electric and Water Meter cases, you need the starting value and then an ending value some time later ( keep track of the time between your readings – a month is a good time frame ).
9) Simply take the Final Number you record and subtract the First Number you record. This Net Amount Used is what you consumed.
10) In the case of Electricity it is normally in kilowatt*hours ( kWh ) while in the case of Water it is 100s of cubic feet of water ( Ccf ).
11) Strangely, the Water Company typically charges for Ccf and not gallons. For those who want to know the gallons, take the Ccf number and multiply by 748 as there are 7.48 gallons in a cubic foot )
12) In both cases, one can call or look up on line the average cost per kWh or Ccf and then simply multiply your use by these values respectively to arrive at an estimated cost.

### Cost = Net Amount Used x Rate

13) Note that this cost is estimated, since it does not factor in any type of tier-pay system for cost changes for different amounts, any taxes, fees, and other costs that may appear on the bill.

## Home Project Needs for Painting and its Cost

1) The most basic of calculation is simple enough, first measure the Length ( L ) and Height ( H ) of a wall in a room in the chosen units ( ft or m ) and calculate the Area ( A ). ( **A = L \* H** )
2) To make the process simpler and avoid mistakes round all values up to the next whole number value. Most walls are 8 ft tall for example.
3) Be sure to sum up all of the walls. TA = total area ( **TA = Sum of all Areas** )
4) Not Recommended , But For those who like some level of precision, in the case of inches divide the inches past the feet measure by 12 and tack this decimal value onto your feet measure, instead of the rounded up values used.
5) For still greater precision, you can subtract out non-painted areas, such as windows and doors ( A = L\*W ) if you like.
6) Whatever the case, take the final number TA and divide by 300. This gives the number of gallons needed for rough, textured, or unpainted wallboard.
7) If it is smooth walls instead divide TA by 350.
8) Note that these are estimates. Always overestimate and round up. For example you are not going to buy 3.3 gallons, buy 4.
9) The final calculation is cost, simply take the Cost per gallon on C scale place over 1 on the D scale and read along the D scale to the needed number of gallons. The Cost is found on the C scale above.

## Home Project Needs for Wallpapering and its Cost

1) The most basic of calculation is simple enough, first measure the Length ( L ) and Height ( H ) of a wall in a room in the chosen units ( ft or m ) and calculate the Area ( A ). ( A = L \* H )
2) To make the process simpler and avoid mistakes round all values up to the next whole number value. Most walls are 8 ft tall for example.
3) Be sure to sum up all of the walls. TA = total area ( TA = Sum of all Areas ) This is the Wallpapering Area.
4) For still greater precision, you can subtract out non-papered areas, such as windows and doors ( A = L\*W ) if you like, instead of rounding up. Subtract the total of the non-papered areas from the papered ones
5) Use the chart below to find the Usable Yield Value. This is divided into the Wallpapering Area to determine the Number of Single Rolls Needed.

$$\textbf{No. of Single Rolls} = \frac{\textbf{Wallpapering Area}}{\textbf{Usable Yield}}$$

6) The final calculation is cost, simply take the Cost per roll on C scale place over 1 on the D scale and read along the D scale to the needed number of rolls. The Cost is found on the C scale above.

| Pattern Repeat ( Drop ) | Usable Yield ( American rolls ) | Usable Yield ( European rolls ) |
|---|---|---|
| 0 to 6 in. | 32 sq ft. | 25 sq ft. |
| 7 to 12 in. | 30 sq ft. | 22 sq ft. |
| 13 to 18 in. | 27 sq ft. | 20 sq ft. |
| 19 to 23 in. | 25 sq ft. | 18 sq ft. |

## Home Project Needs for Carpeting and its Cost

1) The most basic of calculation is simple enough, first measure the Length ( L ) and Width ( W ) of the room in the chosen units ( ft or m ) and calculate the Area ( A ). ( A = L * W )
2) To make the process simpler and avoid mistakes round all values up to the next whole number value.
3) *Not Recommended But For those who like some level of precision, in the case of inches divide the inches past the feet measure by 12 and tack this decimal value onto your feet measure, instead of rounding up.*
4) Determine the Number of Square Yards Needed by taking the Total Area ( in square feet so far ) and divide by 9.

$$\textbf{Number of Yards Needed} = \frac{\textbf{Total Area in sq ft}}{\textbf{9}}$$

5) Note that this Calculation is for Vinyl Flooring too! Follow the same procedure noted above down to the cost here below.
6) The final calculation is cost, simply take the Cost per yard on C scale place over 1 on the D scale and read along the D scale to the needed number of yards needed. The Cost is found on the C scale above.

## Determining Recipe Needs through Proportions

1) These directions addresses the question of what to do when a recipe does not match your materials on hand and/or the question of how much is needed when you are changing the scale of the recipe to a larger or smaller yield.
2) This is the type of activity that the slide rule does very well since it is a proportion, which slide rules excel at.
3) The idea to solve this comes from noting that the ratio of the needed amount for a given ingredient in the recipe of a given number of servings will equal
4) The amount you need to have for your batch divided by the number of servings you wish to make!

$$\frac{\textbf{Recipe Requirement for Material}}{\textbf{Recipe Number of Servings}} = \frac{\textbf{The amount of Material Needed}}{\textbf{Number of Servings}}$$

5) This type of logic can be used for any and all variations of the same theme here. Remember to let the Numerator be the C Scale while the Denominator is the D Scale.

6) First place the known values of the Recipe over each other and look along the D Scale to the Number of Servings you wish to make and find the amount of the unknown needed material above. Be sure to employ the proper use of scientific notation when needed.

7) Keep in mind, you can determine how many servings you are making by looking in the opposite direction and going from what you have and looking back on the D Scale at how many servings it will make!

## Basic Conversions :

1) The Slide Rule is the best conversion system around. This is because this operation is similar to the percentage calculation and the recipe calculation above.

2) Here, in this activity we can convert fractions to decimals or vice versa. We can also find equal fractions as well.

3) Also, if we need to convert one unit into another ( inches into feet, feet into yards, ounces into pounds, inches into centimeters, and even complex ones like mph into kph ) this is considered here.

4) In the case of fractions simply set the C Scale as the Numerator and the D Scale as the Denominator. Reading along the C Scale one finds all of the similar ratios.

5) ( For example, if 2 on C is over 3 on D one finds 4 on C over 6 on D, and 6 on C is over 9 on D, et al. –

6) Not only that, but above the D Scale Index we find the decimal equivalent of the fraction, 0.66 )

7) This means for any fraction or decimal we can find the other easily.

8) What is needed is first a ratio of what is known, say one unit over another. ( For example 1 inch = 2.54 cm, 8 ounces = 1 cup, 1 foot = 12 inches )

9) Set up this ratio on the slide rule. It can be in either order where one is the C Scale value and the other is the D Scale value.

10) Now look for the other known value along the line that is known. On the opposite Scale will be the sought out answer.

11) ( For example if Inches are on the C scale and Centimeters are on the D Scale and we wish to know how many centimeters are 4.5 inches, we read along the C Scale to 4.5 and find the answer below on the D Scale ).

12) What this implies is that with a simple list of conversion factors, one can readily perform rapid calculations.

13) Conversions can include currency, one set of units into another, and the like.

14) For Currency conversions : 1 unit ( say the dollar ) equals N units of another currency. Set the Left Unit ( 1 ) over the equivalent on the D Scale. Reading along C is the number of dollars and below is the number of corresponding units in the other currency!

15) Another fact is this : Recall our calculations previously for miles per gallon, average speed, and cost per unit ounce.

16) Notice that each of these is a 3-Variable Formula where :

17)

18) What this means is that if the ratio is known ( X & Y ) then the answer is always found above the D Scale Index ( N ).

19) Also if we know the outcome value ( N ), we can estimate answers more readily.

20) ( For example if our average speed is 45 mph, how long will it take to travel 280 miles. With 45 on C over 1 on D, read along C to 280 and below it on D is the answer : 6.2 hours )

21) This is how any and all 3-Variable equations can be treated!

22) Some of the basic set of 3-Variable functions needed are as follows :

$$\text{Rate} = \frac{\text{Cost}}{\text{\# Items, etc}} \qquad v = \frac{\Delta d}{\Delta t} \qquad a = \frac{\Delta v}{\Delta t} \qquad V = I*R \qquad P = V*I$$

23) We have used the first 2 here in this activity amongst others ( area, etc ) while the last 2 are Ohm's Law concerning voltage, current & resistance, and the final one is electrical power. The middle one is the formula definition of acceleration.

24) Note that there are many more and this is a small list. The Activities have these to learn from and use. Some involve the amount of food one eats and the graphical breakdown of these items. Another Activity calculates the amount of energy used and power exerted in exercising.

25) The key to this is your use of the Slide Rule as the means to connect math to the real world.

26) The Proportion Formula can be used in a number of cases ( see other Activities involving its use ) such as in the height of an object or distance to it.

27) For example if you know your height and measure the height of your shadow on a given sunny day, and measure the height of a shadow of another thing ( tree, light pole, house ) you can determine its height.

$$\frac{\text{Your Height}}{\text{Size of Your Shadow}} = \frac{\text{Height of Object}}{\text{Size of Shadow}}$$

## Home Economics - Simple Interest on a Loan or Savings Account

1) This calculation is a quick look at the Basic Interest Equation and not the more complex one of being compounded daily or with some other time value.

2) Here, the Principal ( P ) is known, and perhaps the Interest Rate ( R ), along with the time period ( T ).

3) From these variables we can find the amount of accrued interest for a total time period ( I ).

4) This is a multiple step problem, so follow the directions.

5) First place P on the C Scale over the Index of the D Scale.

6) Read along the D Scale to R and find the answer to this on the C Scale, which we will call 'V'.

7) Now move 'V' to the D Scale Index and again read along the D Scale to the variable T.

8) Above T find the Answer I now on the C Scale.

9) Be sure to rearrange the formula as needed if solving for some other variable.

$$I = P*R*T$$

## More Complex Loan Equation Calculations

1) This next excursion is the maximum use of math for this activity and requires concentration. Here we are going to examine the compounded interest rate on a loan or savings account.
2) First look at the equation :

$$A = P*(1+\frac{r}{n})^{n*t}$$

3) Here : P = Principal value, r = Annual nominal interest rate expressed as a decimal value, t = the number of years, n = number of times the interest is compounded per year, and A = Total Amount
4) All of the variables are known here except A. We want to determine the total amount owed or earned from the Principal at these given rates and conditions.
5) First compute with your slide rule r / n by using 'r' on the C Scale over 'n' on the D Scale. Call this X. Jot it down for reference.
6) Add 1 to X and multiply this by P on the slide rule.
7) For example place P on the C Scale over the D Scale Index and read along the D Scale to ( 1+X ) and find the answer, Z, on the C Scale above.
8) Next multiply n*t on the slide rule. Call this M and jot it down.
9) Rewrite Z in Scientific Notation.
10) Reading Z ( see step 7 ) on the D Scale, find the log (Z) on the L Scale. Recognize that this is the characteristic and the mantissa is the exponent from the Scientific Notation value. This new value is Q.
11) Now multiply this value, Q, by M ( step 8 ) and find the value W.
12) Look up the characteristic ( all of the numbers past the decimal ) of W on the L Scale and find its corresponding value on the D Scale.
13) The decimal place is determined by the mantissa of W ( the numeric value before the decimal ).
14) With proper placement of the decimal, the value read on the D Scale is the Total Amount, A.
15) This takes time to do and to understand, but with patience, practice, and determination, you can succeed. As Slide Rulers say, Keep On Computing!

# Activity #2
## Weather Monitoring and Measurement
Grade Level : Middle School
Math Level : Calculating

Weather, simply put, is the day-to-day conditions of the atmosphere. But, it is clearly not as simple as that as anyone looking at a weather map and listening to a weather forecast can consider. In fact, it is a very complex science that stems from many measurements, their changes in time, and each in turn is a mixture of different factors affecting them.

The importance of Weather cannot be emphasized enough. It is the determining factor for appropriate outer wear on a given day, whether or not we have a sunny parade or a rained-out game, whether to dress warmly for cold weather or vice versa, and leads to travel considerations not just for vacations but for work, school, and the movement of goods. Weather damages can be costly and devastating which includes people's lives, livelihood, homes, and communities. Weather affects how we live, dress, even work, and effects everyday safety and costs due to its effects on our structures over time. Today, we can also look to weather to determine if there are cost effective means of energy generation, such as places for solar cells and other solar energy equipment, wind turbines, wave turbines, and the like. Also long term studies of the weather can help indicate changes occurring to the system and our connection to them, such as global warming.

Weather, historically, has only been monitored for approximately the last 2 centuries and the first noted 'meteorologist' of fame is Ben Franklin. He noted the connections of the movements of storm fronts, took measurements and kept records of temperatures of the air and waters in the Atlantic Ocean ( here he found the gulf stream, for example ), and made a famous connection of lightning as a static electric build up and discharge from active storms.

The main place of action is the Atmosphere. Earth's atmosphere is composed of 78% nitrogen ( $N_2$ ), 21% oxygen ( $O_2$ ) and 1% a number of other gases, such as carbon dioxide ( $CO_2$ ). The atmosphere is not uniform throughout, and its pressure is greatest at its lowest layers ( nearest the surface ) and decreases with altitude ( hence the need for pressurized airplane cabins and air masks for high altitude pilots as well as mountain climbers ).

The layer we live in on the surface and extending up to about 18 km is the **Troposphere**. This is the layer in which weather takes place and has the most of the water vapor which can become various forms of precipitation along with clouds formed in this layer.

Briefly, the second layer is the **Stratosphere** and extends form 18 km to 55 km. Here there is no weather to speak of and airplanes travel here. The highest part of this layer has the ozone ( $O_3$ ) which is useful for blocking UV light from the Sun.

Above the Stratosphere is the **Mesosphere** ( 55 km to 80 km ) and above that is the **Thermosphere** ( 85 km to 100 km ). In the Mesosphere the temperature continues to drop rapidly as does the pressure as well. Though the pressure continues to drop into the Thermosphere, the temperature climbs instead of continuing to fall. Above that we are entering space and this edge is referred to as the Exosphere.

The *primary energy engine of the atmosphere is the Sun*. Its radiant energy is mostly light, but there are also other electromagnetic wave energies, such as infrared, ultraviolet, and the like, only in much smaller percentages. Of 100% of the Sun's energy reaching the Earth, 20% is absorbed directly by gases, clouds, and dust in the atmosphere, 50% is absorbed by the Earth's surface while 30% is reflected back into space off of dust and clouds.

In the case of the 50% of the energy reaching the Earth's surface, the surface is not heated evenly ( since 75% of the Earth's surface is water covered and the remaining 25% of ground has altitude variation along with curvature hence varying angles for the Sun striking it ).

As with all energy, the sunlight reaches the Earth as radiation and once it interacts with matter it is then redistributed in the form of **radiation**, or moves molecules by means of **conduction** or **convection**. Some of the molecules in the atmosphere are warmed directly by radiation from the Sun, though this is a very low percentage. Most of the Sun's energy strikes matter such as the surface and warms it, which is turns warms the atmosphere by being turned into thermal radiation ( infrared ). The Earth's surface gives off this thermal radiation which in this longer wavelength form can be absorbed by the atmosphere. Also, the atmosphere closest to the surface when the surface is heated by the radiation conveys the energy to the gas molecules contacting the surface by conduction. Convection currents form from differences in air temperatures. Warmer air masses expand and rise, which in turn begin to cool, while cooler air masses are denser and so will sink to the surface. ( The old adage, hot air rises – cool or cold air sinks ).

We measure the energy of the air as **Temperature**. Temperature, as you might recall is the average kinetic energy of the molecules making up a material. Here we can measure it in Fahrenheit or Celsius typically. Temperature is measured by a thermometer which in the traditional form is the reaction of a liquid in a glass enclosed tube. Later ones involve metals that react to temperature changes, such as thermostats. The range of temperatures of the air on Earth is considerable on the Earth, as illustrated by the polar ice caps and the warm tropical climate areas near the equator ( from -20$^\circ$F normally at the poles to up to over 100$^\circ$ F at the equator ). These are primarily due to the amount of exposure of sunlight to these areas. Also during the day, there are daily highs and lows for the temperature and these can be considerable too, such as in desert regions.

Winds technically occur both vertically and horizontally, though we often refer mostly to the winds that move horizontally across the Earth's surface. These winds are the result in pressure differences. **Pressure**, is force per unit area, and is essentially the weight of the atmosphere on the surface of the Earth ( On average it is 14.7 pounds per square inch which is called 1 atmosphere and is 101.3 kPa in the metric system ). The first influence on pressure by other factors is temperature, and the higher the temperature typically the lower the pressure. This is due to the fact that the warmer air molecules are more energetic ( i.e. have higher average kinetic energy values – definition of temperature ) so there is greater space between the molecules hence lower pressure.

One might think that the air would have greater pressure with water vapor in it, yet this is not the case typically. The greater the water vapor in the air, the lower the pressure. Notice this the next time a storm front passes through. On high pressure days, the skies are often clear – note they can be warm or cool. On low pressure days, the air is humid and in a rain storm the pressure is not only decreasing but lower than average.

Also with altitude, air pressure decreases. The Earth's atmosphere is like a stack of piled up pillows one atop the other. The one at the bottom has a much higher density since the molecules are pressed together by the weight of all of the other pillows atop of it. The same is true for Earth's atmosphere, which has its greatest pressure and density near the surface and a decrease as one rises in altitude.

Air pressure is measured with a device called a **barometer**. The earliest ones used mercury in a closed inverted tube into a cup of mercury. Air pressure would hold a column of mercury at a given level. In fact the idea of standard pressure, 1 atmosphere, was originally considered this way since at this pressure it is 760 mm tall or 760 torr as it was called ( named after Evangelista Torricelli with the creation and use of a mercury barometer in 1643 ). Note that not all barometers use liquid, such as aneroid types with metals that bend according to air pressure.

Back on the topic of winds : In general, warmed air rises vertically and creates a region of low pressure. Higher pressure cooler air masses nearby move horizontally towards this region of low pressure naturally, hence **winds**. When the air moves up or down these are referred to as updrafts or downdrafts respectively. When moving in large masses in this up-and-down manner they are called **air currents**. These are the result of uneven heating of the Earth's surface by the Sun. As noted, winds are mostly generated by a difference in Pressure, hence the reason it is monitored and shown on weather maps. High pressure regions rotate clockwise ( north of the equator ), while Low pressure regions rotate counter-clockwise ( north of the equator and for each, opposite rotation for them south of the equator – known as the coriolis effect ).

The greatest influence on local winds after all of the large effects stem from the topography of the location. There might be **Land and Sea breezes**. Sind the land heats up faster during the day, it will have lower pressure and the high pressure air masses over the water will come in off the sea. At night, the land cools off faster ( those that heat up faster, the faster they cool off too ) and hence will have higher pressure, so the high pressure air mass here will come from the land and move out to the sea which has lower pressure. Similarly there are Mountain and Valley Breezes.

Wind speed is measured with a device called an **Anemometer** ( a device with spinning cups where the speed of the whirling cups is determined to find the speed of the air ) and a Wind Vane indicates direction. When it is 3 kph NW, for example, it is moving at 3 kilometers per hour from the north-west. In the present there are major wind turbines being used to take advantage of the differences in Earth's terrain and heating so as to take advantages of winds that occur in places to generate electricity naturally.

The water in the air is one of the greatest factors controlling the outcome of weather. It is estimated that there is as much as 14 million tons of water vapor in the air at any one time. Of course, it is not equally distributed, as can be noted when comparing very arid deserts to tropical areas with persistent rain. The water comes directly from the surface of the Earth, often from the oceans and through the process of evaporation. **Evaporation** is the process of a liquid turning into a gas. This is when molecules are moving fast enough due to having enough energy to break the bonds that hold them in a liquid state and they become turned into a gaseous state. The Sun is the primary energy source for this, much like for wind.

**Humidity** is a measure of the amount of water vapor in the air. It can go from 0% to 100%. Very low values means there is very little water vapor in the air, while high percentages means a lot of water vapor in the air. Though the term humidity is used in the weather, there are actually different measures of it. One is the **Specific Humidity**. It is the number of grams of water vapor in 1 kg of air. The one we are more aware of is the **Relative Humidity**. It is the amount of water vapor in the air compared with the amount of water vapor that the air can hold at capacity at that temperature ( hence it is a relative percentage of that value % ) Air at capacity (is the amount of matter that something can hold ) is said to be saturated. A **Psychrometer** is common tool used to measure the relative humidity and is often done with dry and wet bulb temperature measures. The dry bulb temp measures the air temperature while the wet bulb thermometer measures the temp relative to the rate of cooling of that thermometer. If it is dry, it cools faster since the water will evaporate from it more quickly, while it cools more slowly in higher humidity times since the water cannot evaporate as readily. ( see table of values which we can use in the activity ).

Another weather term is the Dew Point. The **Dew Point** is the temperature to which air must be cooled to reach saturation. This is the temperature at which the water vapor in the air will reach **condensation** ( the process of changing from a gas into a liquid ). Naturally warm air will hold more water than cool air, so when any given air mass cools, at a given point, the dew point, the water will condense out. This is why we have dew in the morning – the ground at night cools faster than the air above it, hence dew can form on it. If it is cold enough, then it is not dew, but frost ( since water freezes at 0° C, 32° F ).

Condensation does not only take place on the ground, however. It can happen right in the air, when the conditions are right. This is what we commonly call clouds. What of the Clouds, then ? Clouds form from the condensation of water droplets and ice crystals around bits of dust in the air. There are a few types of clouds and their shape determines what it is and how it formed.

Clouds for the most part tend to be in the lowest layers of the atmosphere, the Troposphere. Some do go higher, though. The 3 basic families of clouds ( excluding extreme weather system types ) are : **cumulus clouds, stratus clouds, and cirrus clouds**. The **cumulus clouds** tend to be big and puffy and form from rising currents of warm air that cool as they expand to higher altitudes. The bottoms of them tend to appear rather flat. **Stratus Clouds** can form at similar altitudes as cumulus clouds and are more sheet-like in appearance. Each of these types can be at regular altitudes for clouds or a bit higher and have the prefix attached of alto-. Still higher clouds, called **Cirrus**, are ones that are very thin and wispy in appearance and normally made of ice crystals ( rather than water droplets ) are cirrus clouds at are found above 10,000 km in altitude.

**Fog**, itself, is merely a cloud that is at the Earth's surface. This happens when the ground cools quickly at night and the air above it is cooled to the dew point. It also forms over bodies of water since they tend to be warmer and a cool air mass may move in over them and the water then condenses out.

Extremely large clouds are called **Cumulonimbus Clouds** and rise from very low altitudes and reach to very high altitudes in stature and structure. They can be as tall or taller than Mt. Everest. At their peak, they have what is often referred to as an 'anvil' appearance since they are flattened out. They are typically associated with strong and severe storm fronts and are thunderstorm systems themselves.

Any type of water that falls from the sky is **Precipitation.** Depending on its state ( liquid or solid ), it may have a particular name. Liquid water is simply called Rain, for example. There are 4 main types of precipitation : **Rain, Snow, Sleet, and Hail**.

In the cases of Rain, Snow, and Sleet, these are droplets of water or ice crystals that are suspended in the air. Since they are moving around they can join together and become larger. Once too heave to be held aloft by the air current forces, gravity wins out and the droplet or flake falls to the Earth. Most are considered to have originally formed from condensation of water ( as a liquid or even small solid ) that forms around small dust particles due to collisions. The temperature determines its form. It also shows how the old adage comes about, 'no two snowflakes are alike'. Each has its own unique path after its own origin of a small ice crystal ( often associated with some bit of dust, but not always ) and it has its own cycle of motion, which is highly random in the cloud. It is buffeted by winds, collided with by various other particles and subject to changes in pressure and temperature along its path. Hence its own unique form. If there are layers of cold air when the rain is falling, it may be cold enough for the water droplet to freeze and is then referred to as **Sleet**. Ice Storms can come from this as well as rain that hasn't frozen but reaches the ground and then freezes. In the case of **Hail**, it is essentially lumps of ice that develop in the clouds themselves. Initially they are small ice crystals ( like snow was originally ) but they continuously rise and fall in the cloud system due to air currents and in the process they accumulate more and more ice layers. Like other forms of precipitation, when too heavy, they fall from the sky — but unlike others they have a great and potentially damaging mass to them.

The air in the atmosphere is similar throughout, but there are things we refer to as **Air Masses**. These are large volumes of air which have approximately the same characteristics throughout, such as temperature, moisture content, and pressure. These typically form over large regions ( usually 1,000 km across ). There are 4 main types of Air Masses : **Continental Polar, Maritime Polar, Continental Tropical, and Maritime Tropical**. Essentially they are warm or cold and form over land or water.

Since there are different air masses, there must be 'boundaries' between them. This area is called a **Front**. There are essentially two basic types of Fronts : **Warm and Cold**. Technically defined the front is the boundary between two air masses of different densities. Most air masses travel with the overall rotation of the Earth, hence winds in the continental US move from west to east.

A **Cold Front** is the forward edge of a cold air mass. This front is formed when a cold air mass pushes underneath a warm air mass, since it is denser than the warm air mass ( which is typically rising comparatively ). ( The density issue is connected to Archimedes Principle where objects of lower density float in materials of higher density ). When the cold front moves in, some gusting winds can happen ( due to differences not just in density but also pressure ). The now more rapidly rising warm air will form clouds, hence cloudy skies, and typically rain or some form of precipitation can occur.

A **Warm Front** is the opposite, or it is the edge of a warm air mass behind it. The moving warm air mass will move up and over the cold air mass ahead of it, since it is forced up due to differences in density. On average, the warm air moves more slowly and rises in a sort of a gentle slope over the cold air mass. As it rises, the air mass cools and forms high altitude, typically cirrus clouds

Two other Fronts are possible : **Stationary and Occluded**. In a Stationary Front, there is very little change in weather as each of the warm and cold air masses remain relatively in the same place over a period of time. The other type, the Occluded Front is the most complex and occurs when a warm air mass is nestled between two cold air masses. In time, the cold air masses will usually force aloft the warm air mass ( making it occluded from the ground ) and join. The warm air mass cools and will turn cloudy and often have some form of precipitation.

Severe Weather comes in three basic forms : **Thunderstorms, Tornadoes,, and Hurricanes**. A **Thunderstorm** is one that has heavy rains, along with lightning and thunder and often has strong winds. They come from cumulonimbus clouds. Lightning occurs due to the separation and build up of charges ( positive and negative ) through the cloud system and there is more lightning between clouds then the ground, but can occur repeatedly. **Lightning** happens when the charge build up is sufficient to overcome the resistance of the air between the items and a discharge occurs. This in turn heats the air, which expands explosively and creates a compressed sound wave in the air ( the boom of thunder ). Note also that lightning can not only go from the cloud to the ground but from the ground to the cloud. It is the difference in charge which sets up a potential difference that can bring about a discharge. A **Tornado** is a funnel-shaped cloud that is rotating. They can be very small to very large in size. They are often associated with both thunderstorms and hurricanes. If they occur over water, they are referred to as waterspouts. The funnel is a very low pressure region, hence in the area of touchdown, the surrounding higher pressure as opposed to the low pressure region of the tornado creates extreme winds ( it moves naturally from high to low pressure ) and we may refer to this as acting like a vacuum with suction, but know, in science, it does not suck. A **Hurricane** is a tropical storm that forms over water, the ocean, and is a rotating low pressure region. A Hurricane can contain numerous thunderstorms in it and can become very large. It develops over warm waters ( where the energy of the system comes from ). Due to low pressure and excessive winds, it will build up a pile of water that ends up being pushed on land called the storm surge.

With all weather as time as gone by there are more and more tools and tools of greater and greater accuracy to map the details of characteristics such as : temperature, wind speed, pressure, and the like. Data is recorded locally, and over large regions with satellites. From this a **weather map** may be generated. On it there will be High and Low Pressure regions. **Isobars** are lines that connect points of equal pressure, while **isotherms** connect points of equal temperature. A cold front is often represented by a dark blue line with triangles on it, while a warm front is a dark red line with half circles on it. A stationary front is an alternating pattern of these of red and blue in opposite directions, while the occluded front is purple and alternates triangle then half circle pointing in the same direction. This measured information in the present coupled with years of data are used to generate computer-projected models to predict the course of systems and their changes over time. In the present, most forecasters can predict with some reliability 1-3 days with some confidence ( though most of us may disagree ). With any long range forecast, the more days that there are, the lower the reliability.

Though weather is the day-to-day conditions, **Climate** is the overall average conditions from year to year and includes the averages over extended periods of time. Not only is climate studied through measurements of the weather, but climatologists examine the conditions long ago by examining such things as trapped air bubbles in ice ( looking for the presence and amount of $CO_2$ in the air), studying tree rings ( for wet versus dry seasonal variations ), and the like from nature to indicate changes through time of the climate. This is an area of concern today for many reasons.

Our Activity is where we are the weather-people. Living where we do, when we do, we will monitor various things, such as temperatures ( daily highs and lows ), the amounts and types of precipitation, wind speed and direction, humidity values ( best to determine it from a table and using a wet & dry bulb system ). It is recommended to use as simple a set of instruments as possible due to cost as well as learning how to use various tools with scales on them ( as opposed to electronic devices ).  The key is parental support and help in this. Always act in a safe manner concerning weather as well as using the tools correctly and safely as well.

**Purpose :** To monitor changes in the weather to do basic calculations with the values to determine average high and low temperatures with time, average precipitation for a time period, average wind speed, wind chill factor or heat index, compare one's measurements to local weather readings, and to possibly predict weather in the immediate future.

**Materials :**

> Weather Measuring Tools – these should include ( outdoor capable ) :
> Minimum – Maximum Thermometer,
> Barometer ( air pressure ),
> Mason Hygrometer ( relative humidity determination ),
> OR Relative Humidity Gauge,
> All Weather Rain Gauge ( rain, snow, sleet fall ),
> Anemometer ( wind speed gauge ),
> Outdoor wood or plastic box for instruments,
> Record Book for data,
> Slide Rule

\*\* NOTE : many of the above items can be purchased separately or collectively. The price range is considerable, but most items are below $40 individually. It is recommended to not use digital since learning to read thermometers and gauges with pointing needles at a scale is a useful skill. Also there are websites that use basic materials to obtain estimated values of weather factors, such as using paper cups on a spinning spool and the calculation of its rate of spinning to arrive at a value for wind speed. Some instruments, such as the Mason Hygrometer will yield Dry & Wet bulb temperatures – so this requires refilling with water for the wet bulb and the use of a table to determine to relative humidity from the device's readings.

**Set Up :**

> These outdoor instruments should be in a box where the thermometer is not in direct contact with surfaces that can convey heat to the thermometer so as to affect its readings. Note that the rain gauge is the one tool that is outside the box and must be checked regularly to avoid the loss of liquid due to evaporation.

## Procedure :

1) Set up the Weather Station so that it shields the main instruments and allows the rain gauge to be outside of it. For all instruments be sure of proper knowledge of use, reading, and set up so as to minimize external influences on the data and maximize their effectiveness.

2) Choose a time frame for data to be recorded. To be effective it should be as long as possible. For those doing this on their own, doing this for a month during each of the major seasons ( Spring, Summer, Autumn, and Winter ) would be good. In school situations, a two week span in a given semester will be good minimally.

3) Each of the instruments will have their own pattern of readings. For example, the Minimum – Maximum Thermometer only needs one reading per day, usually in the morning. Other instruments, such as the Mason Hygrometer can be read at more than one time per day and each reading will result in a Relative Humidity value for that time of day. In this case, twice a day is useful and each of these times should be separated by 6 or more hours.

4) When it comes to some instruments be sure to reset them as needed, such as the Minimum – Maximum Thermometer.

5) Create a Weather Log of your readings. An average day will have these readings : Minimum Temperature, Maximum Temperature, Relative Humidity, Wind Speed, Wind Pressure, and Precipitation for that day.

6) It is best to use the Dry and Wet bulb method and the table provided to determine relative humidity, rather than a device that yields this value. For more calculation fun, do both and compare the differences of these values.

7) Always leave space for comments, such as rapid changes in the weather, estimated amount of cloud cover for that day, and the like.

8) For comparison use the daily news ( TV, radio, internet ) for your area's values for some of this data for comparison and record it as well. Other research can reveal average values for comparison.

9) Note and Reflect on Readings to see if there are patterns. See if this allows you to predict a possible range of values for the following day or two. ( For example, watching the temperature over several days, what value may it be tomorrow? ) Consider hypotheses too. Do rapid changes in the pressure precede weather changes ( such as precipitation or high wind speeds )?

10) Calculate the Averages for the following : Average Low Temperature, Average High Temperature, Average Daily Temperature, Average Wind Speed, Average Relative Humidity, and Average Precipitation for a given time period.

11) More thorough analysis can examine the possibility of extreme differences in temperatures as compared to the average temperature. Here it is sample variance and sample standard deviation.

12) Make a graph of results : Daily High & Low Temperatures ( y-axis ) versus the day ( x-axis ) as line graphs.

13) Use the data to compute either the Wind Chill Temperature or Heat Index Temperature ( depending on the time of year, of course ).

14) Use the data to calculate the Dew Point Temperature as well and compare to results noted in the media for that day.
15) Other Calculations that can be done : Converting Pressure from the units your device measures to other units : psi to atm to mbar to mm Hg, : wind speed from mph to kph, : temperature in °F to °C, and others you may consider.
16) Be sure to use your Slide Rule for any and all calculations. Note that the formulae used here are approximations in some cases (more exact ones can be found on the internet ).
17) For Slide Rule calculations know how to use primarily the C & D Scales for multiplication and division.
18) To calculate formula, such as the wind chill factor note to use the L Scale. Look up the value of V, the wind speed in mph, on the L Scale as read from the D Scale. Multiply this by 0.16. Now look up this resultant value on the L Scale to find the outcome on the D Scale ( noting that any value in the ones place is a power of the outcome – i.e. characteristic while the decimal value, the mantissa – is the corresponding log value of the number sought ) to put in the formula.
19) For the formula for Dew Point, you need two things : the first is that you are using both a 'dry bulb' and 'wet bulb' to find temperatures. Next use these to find their difference and look to the chart provided to find the Relative Humidity ( RH ). Then to find the Dew Point Temp use the natural log of the relative humidity value expressed as a decimal value. ( example 50% = 0.50 ). In the provided formula. To determine this with a 9-scale slide rule, take the log of the relative humidity value and divide it by the log of the natural log base, e, which can be approximated as 2.72.

## Data :

Personally recorded data from instruments :

- Note : Either the Humidity is measured or calculated

| Date | Temp (high) | Temp (low) | Wind Speed | Pressure | Humidity | Precipitation |
|------|-------------|------------|------------|----------|----------|---------------|
|      | Units       | units      | units      | units    | Units    |               |
|      |             |            |            |          |          |               |
|      |             |            |            |          |          |               |

| Date | Dry Bulb Temp | Wet Bulb Temp |
|------|---------------|---------------|
|      | Unit          | Units         |
|      |               |               |
|      |               |               |

Data recorded from news source
- Note : 'Other' can be descriptions of conditions – cloudy, et al and/or changes that occur during the day

| Date | Temp (high) | Temp (low) | Humidity | Wind Speed | Dew Point | Pressure | Precip. | Other |
|------|-------------|------------|----------|------------|-----------|----------|---------|-------|
|      |             |            |          |            |           |          |         |       |
|      |             |            |          |            |           |          |         |       |

## Calculations :

Be sure to use your Slide Rule !

## Average Value Formula

for high temp, low temp, ave. daily temp, ave. wind speed, also can be done for precipitation

$$\text{Average Value} = \frac{\text{Sum of Measurements ( } \Sigma m \text{ )}}{\text{Number of Measurements (N )}}$$

$$\textbf{Ave} = \frac{\Sigma m}{N}$$

$$°F = 1.8 * °C + 32°$$

$$°C = \frac{(°F - 32°)}{1.8}$$

## Heat Index Temperature Formula ( HI ) :

$$HI \sim -42.4 + (2.05)*T + (10.1)*R - (0.225)*T*R$$

T : dry bulb or temp at that time ( °F )
R : relative humidity ( integer % value )

## Wind Chill Index Formula ( WCT ) :

$$WCT \sim 35.7 + (0.622)*T - 35.6*V^{0.16} + 0.428*T*V^{0.16}$$

T : dry bulb or temp at that time ( °F )
V : wind speed in mph ( miles per hour )

### Dew Point Temperature ( $T_d$ ) :

More Advanced Formula :

$$T_d = \frac{a*T°}{b+T°} + \ln(RH)$$

a = 17.3   &  b = 238°C
T° : Temperature in °C for the dry bulb temperature

( Useful for Temps 0°C – 60°C, RH : 1% to 100%, and will work for outcomes of Dew Point Temperatures of 0°C to 50°C )

$$\ln(RH) = \frac{\log(RH)}{\log(e)}$$

e = 2.72

Basic Formula :

$$T_d = T_c - \frac{1-RH}{0.05}$$

RH : relative humidity ( % expressed as a decimal )
$T_c$ : dry bulb temperature in °C

Statistical Examination of Data :
n = is the number of data values, $^-$ is the mean

Sample Variance

$$s^2 = \frac{\Sigma(x- x)^2}{n-1}$$

Sample Standard Deviation

$$s = ( \text{Sum of } (x - x_{ave})^2 / ( n-1) )^{1/2}$$

From this theorem,
- At least 75% of the data fall in the interval $^-$ ± 2s
- At least 88.9% of the data fall in the interval $^-$ ± 3s
- At least 93.8% of the data fall in the interval $^-$ ± 4s

Any data point beyond 3 standard deviations is considered an outlier.

## Conversions :

Wind Speed :
1.61 kph = 1 mph

Pressure :
1 ATM = 760 mm Hg = 14.7 psi = 101.3 kPa = 1013 mbar

Measurements :
1 inch = 2.54 cm
1 hr = 60 min. = 3600s
1 mi. = 5,280 ft = 1.62 km
1 m = 100 cm.

## Useful Data Table
## ( Wet & Dry Bulb Temp Determining Relative Humidity ):

Chart for RH from Wet & Dry Bulbs

The Left-hand column is the Dry Bulb Temperature in Celsius

The Columns running from 0 to 15 is the result of the difference in the Dry Bulb Temperature and the Wet Bulb Temperature in Celsius

The Table shows the results for Relative Humidity as percentage ( % )

| | 0 | 1 | 2 | 3 | 4 | 5 | 6 | 7 | 8 | 9 | 10 | 11 | 12 | 13 | 14 | 15 |
|---|---|---|---|---|---|---|---|---|---|---|---|---|---|---|---|---|
| -20 | 100 | 28 | | | | | | | | | | | | | | |
| -18 | 100 | 40 | | | | | | | | | | | | | | |
| -16 | 100 | 48 | | | | | | | | | | | | | | |
| -14 | 100 | 55 | 11 | | | | | | | | | | | | | |
| -12 | 100 | 61 | 23 | | | | | | | | | | | | | |
| -10 | 100 | 66 | 33 | | | | | | | | | | | | | |
| -8 | 100 | 71 | 41 | 13 | | | | | | | | | | | | |
| -6 | 100 | 73 | 48 | 20 | | | | | | | | | | | | |
| -4 | 100 | 77 | 54 | 32 | 11 | | | | | | | | | | | |
| -2 | 100 | 79 | 58 | 37 | 20 | 1 | | | | | | | | | | |
| 0 | 100 | 81 | 63 | 45 | 28 | 11 | | | | | | | | | | |
| 2 | 100 | 83 | 67 | 51 | 36 | 20 | 6 | | | | | | | | | |
| 4 | 100 | 85 | 70 | 56 | 42 | 27 | 14 | | | | | | | | | |
| 6 | 100 | 86 | 72 | 59 | 46 | 35 | 22 | 10 | | | | | | | | |
| 8 | 100 | 87 | 74 | 62 | 51 | 39 | 28 | 17 | 6 | | | | | | | |
| 10 | 100 | 88 | 76 | 65 | 54 | 43 | 33 | 24 | 13 | 4 | | | | | | |
| 12 | 100 | 88 | 78 | 67 | 57 | 48 | 38 | 28 | 19 | 10 | 2 | | | | | |
| 14 | 100 | 89 | 79 | 69 | 60 | 50 | 41 | 33 | 25 | 16 | 8 | 1 | | | | |
| 16 | 100 | 90 | 80 | 71 | 62 | 54 | 45 | 37 | 29 | 21 | 14 | 7 | 1 | | | |
| 18 | 100 | 91 | 81 | 72 | 64 | 56 | 48 | 40 | 33 | 26 | 19 | 12 | 6 | | | |
| 20 | 100 | 91 | 82 | 74 | 66 | 58 | 51 | 44 | 36 | 30 | 23 | 17 | 11 | 5 | | |
| 22 | 100 | 92 | 83 | 75 | 68 | 60 | 53 | 46 | 40 | 33 | 27 | 21 | 15 | 10 | 4 | |
| 24 | 100 | 92 | 84 | 76 | 69 | 62 | 55 | 49 | 42 | 36 | 30 | 25 | 20 | 14 | 9 | 4 |
| 26 | 100 | 92 | 85 | 77 | 70 | 64 | 57 | 51 | 45 | 39 | 34 | 28 | 23 | 18 | 13 | 9 |
| 28 | 100 | 93 | 86 | 78 | 71 | 65 | 59 | 53 | 47 | 42 | 36 | 31 | 26 | 21 | 17 | 12 |
| 30 | 100 | 93 | 86 | 79 | 72 | 66 | 61 | 55 | 49 | 44 | 39 | 34 | 29 | 25 | 20 | 16 |

## Conclusion :

What do your observations show? How similar or dissimilar are your measurements as compared to local measurements? Do your records and examinations of the data give you the ability to estimate values for the upcoming day or up to 3 days? How different are your temperature values from the average values for the time period covered ( are there any 'outliers' )? Do you speculate that there are any connections between any of the measured phenomena?

## Activity #3
## Height and Altitude Considerations with the Slide Rule
Grade Level : Middle School
Math Level : Calculating

The primary goal of science is the application of math models and formulae to the real world in order to do such things as understand cause-and-effect, make predictions when changes in one variable are made to another variable, and to solve for the unknown in a given situation.

One of the simplest applications is the direct application of math formulae for such things as area and volume as well as similar triangles.

If we start with geometry and a right triangle where the base is 4 units long, the height is 3 units tall, then we can find from the Pythagorean Theorem that the hypotenuse is 5 units long ( $3^2 + 4^2 = 5^2$ ).

Now we change the side from 4 units to 8 units long and we also double the height from 3 units to 6 units. What becomes of the hypotenuse? It too doubles to 10! ( $8^2 + 6^2 = 10^2$ ). Though this illustration only employs whole numbers, any multiplying factor for the hypothetical triangle would work. This is because each of the triangles, the original and the magnified one, are what are called Similar Triangles. Similar Triangles are triangles that have the same angles comprising the triangle and their sides are all proportionate to each other. For example from our triangle example, if we take ¾ on the first triangle it corresponds to 6/8 on the other one. ( also for these triangles, they have the same set of angles : 90°, 36.9°, and 53.1° ).

Also realize that similar triangles do not only apply to right triangles, but to any two triangles that have the prescribe description above applying to them.

What value are these similar triangles? Let's say one of the two triangles have the values known for sides and angles, and from the other one we have the angles ( as needed ) but only one of the two sides – the other is unknown. We could then easily solve for the other side by proportions. Take our example ( let's assume we do not know the corresponding side to 3 in the other triangle ):

$$\frac{3}{4} = \frac{X}{8}$$

Beyond math, this has many applications in the real world. Imagine you hold up a ruler vertically at the end of your arm in your hand. You have measured your arm length ( one side of the small triangle ) and you look at a tree, house, a telephone pole with the ruler and measure its apparent height ( the other side of the small triangle ). Now all you have to do is measure how far away you are standing from the object under consideration ( the distance to the tree ) which gives you the corresponding triangle side length to your arm. The unknown value here is the height of the object in question and is solved like our proportion above.

$$\frac{\text{Apparent Height}}{\text{Arm Length}} = \frac{\text{Actual Height}}{\text{Distance from Object}}$$

The idea of similar triangles can be reduced to one triangle and the use of trigonometric formulae. All one needs to know is the side of a triangle and the angle associated with it. From this one can determine another side. More importantly instead of the triangle(s) being vertical as noted in the early part of the discussion, this is true even if the triangles are on the ground and we can only determine one of the sides! This is simply because there is no special direction in nature. Look directly across a river at an object, such as a tree. Walk parallel to the river a measured distance and look at the tree again. Using a protractor measure the angle from the line you walked to the tree. Here we can use the ratio of two sides of this triangle, the width of the river to the distance walked is the tangent of the angle you have measured.

$$\text{Tan } \Theta = \frac{\text{X-River Width}}{\text{M-Distance walked}}$$

Parallax and Astronomy Use :

One of the most important applications of this idea comes from Astronomy. One of the easiest arguments employed by peoples of long ago to argue that the Earth is not moving was this question : If we are moving, why do not the closer stars appear to move due to our motion as compared to the more distant stars in the sky?

How does this question work? Simple – hold a pencil erect at arm's length in front of you and first look at it with one eye and then the other. Next try this closer to you. The pencil as compared to more distant objects, such as furniture along the wall, appears to move more when closer to you than when farther away.

How this applies to the question of the ancients is this – your eyes are to represent the Earth at two different locations in its orbit. The pencil is a nearby star. If we are moving and we look at the nearby stars, we should see their apparent motion. For centuries this was not detected, because stars are so far away, there apparent motion as it is called is very small. Once the apparent angle is measured, the distance the Earth travels, which is the baseline for the triangle, allows us to determine the distance to the star.

This term in astronomy is called parallax. Defined, parallax is the apparent shift or motion of a relatively close object with respect to more distant objects as the location of the observer changes.

Finally in 1838 the parallax of several stars were discovered. It turned out that even the closest stars have a parallax of less than 1" ( 1 second of arc – Note : 1° has 60 minutes of arc, while each minute has 60 seconds of arc ). The distinction goes to 3 different people of 3 different nations : Friedrich Bessel with 61 Cygni, Friedrich Struve with the star Vega, and Thomas Henderson with alpha Centauri. For those interested, the largest is of course the closest star ( other than our own Sun ) Proxima Centauri at 0.772 arcseconds of parallax, which results in a distance of 1.30 parsecs ( 4.2 light-years ).

Our Activity will be a series of activities and use each of these techniques in turn to determine the size of objects or distance to objects, including a 'star' we make from a ping pong ball to demonstrate the idea of parallax.


**Purpose :** To determine the measure of the height ( altitude or size ) of a distant object by indirect measurements and proportion calculation.

**Purpose :** To determine the measure of the height ( altitude or size ) of a distant object by indirect measurements and definition of tangent calculation.
( First a local one to test the method both ways – poster/painting/mirror/fixture on the wall, then to use on other objects such as the height of a tree, telephone pole, or a house and possibly a launched rocket and its altitude outdoors ).

**Purpose :** To determine the distance to or across a region by using the side of a triangle and the tangent of the angle made by 2 observations of a distant object.

**Purpose :** To measure the distance to an object that represents a star ( ping pong ball ) so as to illustrate the process of Parallax in Astronomy.


## Materials :

- Item to determine height of ( poster on wall, billboard, tree, wall, house, lamp post, telephone pole, et al ),
- protractor,
- string ( possibly dental floss ),
- nut,
- measuring tape ( flexible about 2 m ),
- Measuring tape ( long about 20 m or more ),
- ping pong ball,
- paper clip,
- launch-capable object ( water or air-propelled rocket ),
- slide rule

**Procedure :**

**Activity 1 :** Determination of the height of an object in the home on the wall through proportion method.

1) For the indoor activities take all measurements in centimeters.
2) Tools needed for this activity : Object on Wall ( can be painting, poster, et al ), ruler, regular measuring tape, flexible measuring tape, slide rule
3) Use a measuring tape and from the wall on the floor just below the chosen object to determine its height open and start it there and measure off up to 3m ( this depends on room size ) perpendicular to the wall away from the object.
4) The goal is to be somewhere between 1m and 3m from the object for measurements so that when holding up your hand at that distance the object appears about the size of your hand at most.
5) With ruler and flexible measuring tape in hand pick a spot to stand at along the measuring tape on the floor. Choose a number that will be the distance to the wall ( D ) on the table of data values so that a line from your eye to view the object with to the floor is measured from here to the object.
6) Face the object. Look at it with one eye. With the corresponding hand with both the ruler and flexible tape in it raise it so as to hold the ruler perpendicular to your arm. ( Note : Your arm is parallel to the floor ).
7) Find a comfortable distance to hold the ruler erect at so that the edge can be read ( the centimeter side and the numbers going up vertically ).
8) Once done, reach with the other hand and let the flexible measuring tape unroll as your free hand draws it from your hand with the ruler back to your face in the area of your eye that is used for viewing.
9) Once there, have this hand hold the value on the flexible tape and hold that tape with this hand.
10) Looking at the ruler in the other hand, try to align the bottom of the object with a whole number on the ruler.
11) Keeping this value in mind read it and the value that appears to be the top of the object from the perspective of the ruler from this view.
12) ( Side Note : to optimize the reading in this case align the ruler at the end of your arm with the bottom of the object. This may mean crouching to align ).
13) Record your data. The number on the flexible measuring tape goes in the Arm Length box on the data table ( A ), while the Apparent Height Top ( $R_{Top}$ ) and Apparent Height Bottom ( $R_{Bottom}$ ) go in their respective boxes.
14) Calculate the Apparent Height, which is the difference of the Top and Bottom Values, place in box. ( R )
15) Use the formula to determine the Actual Height of the Object.

16) Do this 3 times from 3 different distances and take the average of your calculated height value for the object.
17) Measure the actual height of the object. Then calculate the percent error of your indirect value to the actual value.
18) Look at the photo for Activity 3 which applies to this case since this one is only a smaller scale of that outdoor one which shows the Ruler Method of Similar Triangles.


**Activity 2 :** Determination of the height of an object in the home on the wall through tangent method.

1) Follow Steps 1 through 4 in Activity 1 Noted above.
2) Stand somewhere between 1m and 3m away from the object along the measuring tape. Write down this value in the table.
3) Materials needed here : protractor, string ( dental floss is best ), lug nut, tape, straw ( thin stirring one is best so long as it is a tube and straight ).
4) As in the first activity it is best to crouch and align your view from the base or bottom of the object on the wall so that your view of the object is a direct line parallel to the floor.
5) Tape the straw along the base of the protractor so that it goes from $0°$ to $180°$. Attach the string to the vertex point of the protractor and have it extend in the direction of the angular measurements and be long enough to tie the lug nut to it ( see photo of Astrolabe ). This is called your astrolabe.
6) Use the protractor device by looking through the straw ( along the bottom of the straw ) first at the bottom of the object where you are directly in line with it so that the nut hanging down is at $90°$.
7) Now slowly rotate the protractor so that you are now looking through the straw at the top of the object.
8) At this point use your other hand to hold the string against the protractor.
9) Read the angle measured here. ( $\Theta_2$ ) – Note this is not the angle used. Take the compliment of this angle ( $90° - \Theta_2$ ) to obtain the needed angle ( $\Theta$ ).
10) Calculate the height of the Object using a slide rule.
11) Do this 3 times like the first activity and average the values determined for the height of the object done by this manner.
12) Using the actual value for measured height, compute the percent error of your indirect measurements.
13) See photo of Tangent Activity in photo of same name.

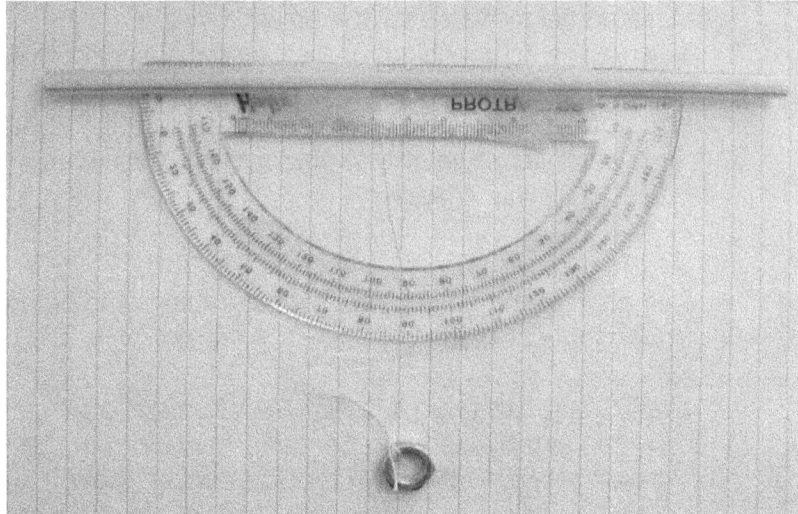

**Activity 3 :** Determination of the height of an object outside ( such as a telephone pole or tree ) through proportion method and the tangent method for comparison.

1) For the outdoor activities take all measurements using the long measuring tape but measure in meter.
2) Tools needed for this activity : Object outside ( telephone pole, light pole, house, flag pole, or tree ), ruler, long measuring tape, flexible measuring tape, slide rule
3) Use a long measuring tape and from the object in question at its base in a direction to stand far enough away so that when holding the ruler erect at arm's length the entire object falls within the scale on the ruler. ( This may be as little as 3m up to 18 m distant ).
4) Record the distance that you are standing at from the object.
5) Now measure the height to your arm which will be extended and parallel to the ground with the erect ruler in your hand. Record this height ( it will be added to the height of the object later on ).
6) Next with ruler and flexible measuring tape in hand at your spot to stand and face the object with arm extended with ruler erect and look at the object with one eye closed aligning the centimeter side of the ruler along the vertical axis of the object starting at its base and going to it's top.
7)  With the corresponding hand with both the ruler and flexible tape in it raise it so as to hold the ruler perpendicular to your arm. ( Note : Your arm is parallel to the ground ).
8) Find a comfortable distance to hold the ruler erect at so that the edge can be read ( the centimeter side and the numbers going up vertically ).

9) Once done, reach with the other hand and let the flexible measuring tape unroll as your free hand draws it from your hand with the ruler back to your face in the area of your eye that is used for viewing.

10) Once there, have this hand hold the value on the flexible tape and hold that tape with this hand.

11) Looking at the ruler in the other hand, try to align the bottom of the object with a whole number on the ruler.

12) Keeping this value in mind read it and the value that appears to be the top of the object from the perspective of the ruler from this view.

13) Record your data. The number on the flexible measuring tape goes in the Arm Length box on the data table ( A ), while the Apparent Height Top ( $R_{Top}$ ) and Apparent Height Bottom ( $R_{Bottom}$ ) go in their respective boxes.

14) Calculate the Apparent Height, which is the difference of the Top and Bottom Values, place in box. ( R )

15) Use the formula to determine the Actual Height of the Object.

16) Redo Activity 2 only as an outdoor exercise using the astrolabe to determine the angular height of the object and knowing the distance from it, you can calculate the unknown height. Use the tangent formula. Look at the Tangent Activity photo.

17) If possible, find records or general information to verify your calculated height as compared to what is known. If impossible, then try the alternate method below which uses shadows and compare results.

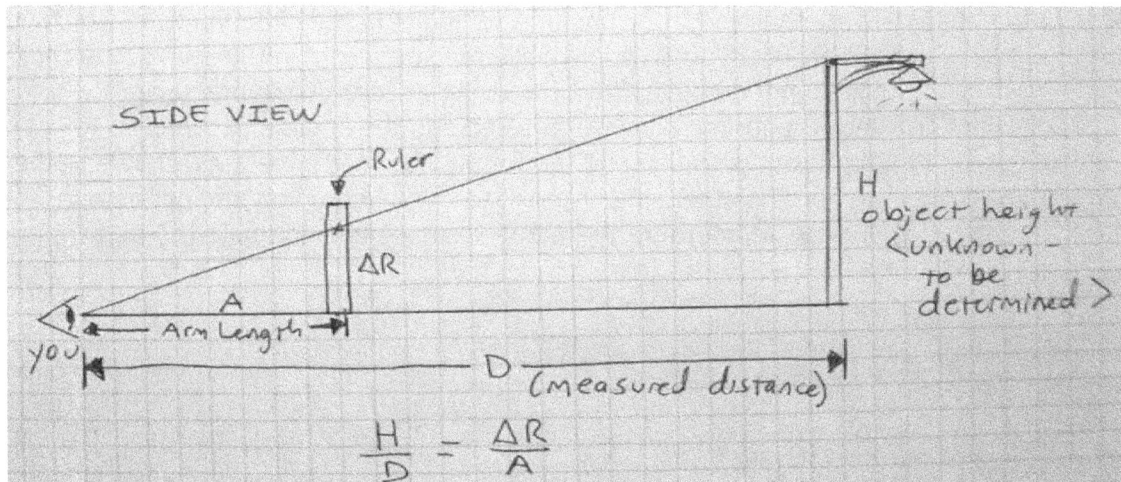

SIDE VIEW

Ruler

$\Delta R$

A — Arm Length

you

H object height <unknown - to be determined>

D (measured distance)

$$\frac{H}{D} = \frac{\Delta R}{A}$$

**Activity 4 :** Determination of the distance across an object outside ( such as the distance across the street ) through tangent method.

1) Use the Astrolabe you constructed in Activity 2 above ( see photo of it ) for this next activity.

2) Stand a good distance ( over 20 ft, perhaps on the order of 60 ft to 100 ft ) but still measurable by your longest measuring tape so that the answer can be determined. ( Measure it later, not now ).

3) A good way to do this activity is to be on one side of a residential ( low traffic volume ) suburban street. The best objects to consider are one of the following on the other side of the street from you : A) light pole, B) fire hydrant, C) tree, or D) telephone pole. Typically some sort of object in a boulevard portion of a lawn is best.

4) Stand directly across from the object so that the street runs between the two of you.

5) Walk in a straight line perpendicular to the object a measured distance ( 10 ft or 3 m ) ( this is 'd' ) so as to make the base of a triangle ( your path is one leg, the other leg is the distance from your starting point to the object and the hypotenuse will be the your view path to the object from where you end up ). ( see photo )

6) Facing the direction from which you came, point the straw back at where you started and have the protractor portion pointed away from the object. Swing the straw in the direction of the object until it is in view, all the time sliding the string and nut so as to be able to measure the angle covered.

7) To emphasize : the string and nut point in the direction from which you came and are marking an angle while the straw with you viewing through it now points at the object. Start the string at zero. Angle measure ☐

8) Record both the distance traveled and the angle measured.

9) Use this information and the formula to determine the distance to the object ( x ) on the other side of the road from where you originally stood.

10) If possible, measure this actual distance and compute the percent error.

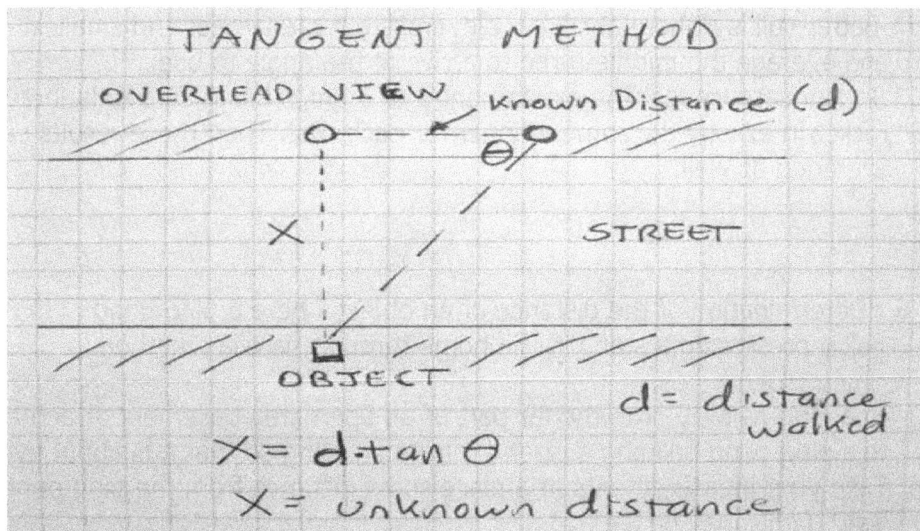

TANGENT METHOD

OVERHEAD VIEW — Known Distance (d)

STREET

OBJECT

$d$ = distance walked

$X = d \cdot \tan \theta$

$X$ = unknown distance

**Activity 5 :** Determination the altitude an object rises to when launched ( such as a water rocket ) through proportion method.

1) This activity can be done in conjunction with the water bottle rocket activity in the book.
2) In any case, one needs either a water-propelled or air-propelled rocket for the activity.
3) Also it is best to have at least 3 friends besides the one launching the rocket ( though the minimum is 2 people – one to launch and one to measure the maximum altitude ).
4) All of those engaged in measuring the maximum altitude need to use the astrolabe constructed in activity 2 above.
5) The rocket launcher readies and activates the rocket. The others performing the measurements are situated in a manner so as to be 3-5 m away from the launch. Whatever the distance, it must be measured and recorded. ( This is 'd' ).
6) When launched, the altitude measurer uses the astrolabe and follows the path of the rocket to its maximum altitude by looking through the straw of the astrolabe and letting the nut and string hand down.
7) Once at the peak the altitude measurer holds the string in place on the protractor so as to measure the angle made.
8) Note that if not measured from zero, you will have to take the reading and subtract it from 90°.
9) The angle and the distance from the launch are used with the tangent formula to determine the altitude of the rocket.
10) If more than 1 is measuring the maximum altitude, have as many as possible measure the maximum height angle simultaneously and average the values for the measured angle.

11) Better still is if there are 5 people, cross out the largest and smallest angles and average the middle three to arrive at the angle to use.

12) In any case, repeat the experiment 2 to 3 times being sure to launch the rocket in exactly the same manner for each trial. Then these results can be averaged as well.

**Activity 6 :** Determination of the distance of an object ( here a ping pong ball representing a star ) in the home through parallax method.

1) In a long hallway or in the far part of an open area basement attach a string to a ping pong ball and suspend it from a ceiling so that it is about eye level.

2) Take your measurements at a place some distance from the ping pong ball 'star'. ( this may be 3+ m ).

3) It is best to have a line to measure from that is perpendicular to a wall that runs in the direction of the 'star' ( see diagram ).

4) Pick two points on this line so that if a line were to be drawn directly from the 'star' to the line it would bisect it the two chosen points would be nearly equal in distance on either side of the midpoint.

5) For the measurements, you use the astrolabe which was constructed in activity 2 above. For measurements you will be doing them horizontally. Hold the astrolabe parallel to the floor.

6) From each point stand so that you can point the straw first at the other point to be measured from and then moved so as to be pointed directly at the 'star' when viewing it.

7) While moving it from the line of observation to the 'star' hold the string and nut so that it sweeps across the protractor and measures the angle from the measurement line to the star from your point of observation.

8) If the angles are not congruent, choose a place so that they become congruent for best results.

9) The measurements to record are the distance between your observation points and the two angles measured.

10) Find the 3$^{rd}$ angle of the triangle from your two angles when added and subtracted from 180°. Now divide this answer by 2.

11) Use the tangent formula to find the distance to the 'star'.

12) Actually measure the distance with a measuring tape and determine the percentage error.

13) See photo below for the basic set up as well as an idea in a sketch.

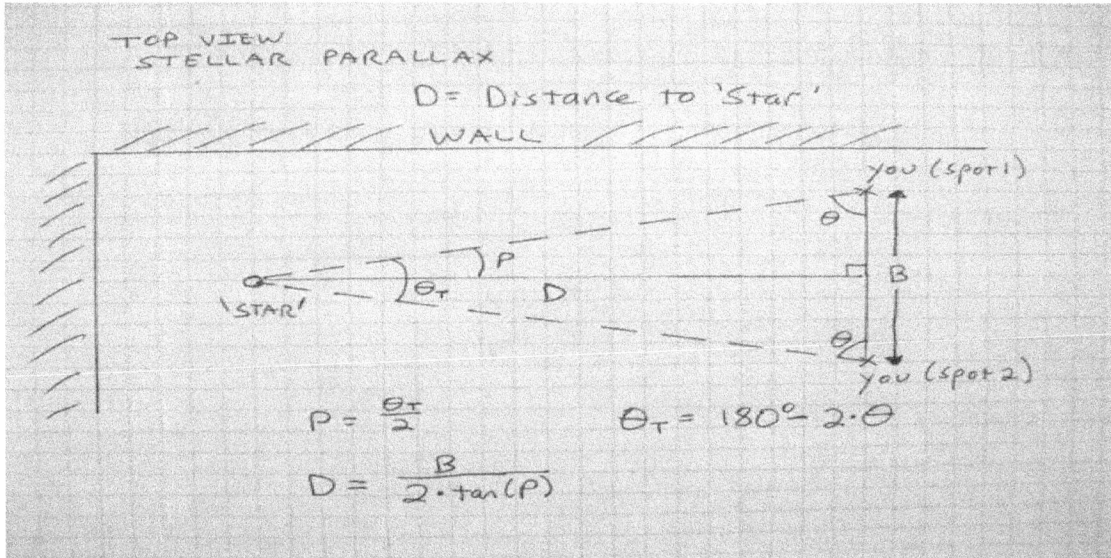

TOP VIEW
STELLAR PARALLAX

$D$ = Distance to 'Star'

WALL

you (spot 1)

'STAR'

$B$

you (spot 2)

$P = \dfrac{\theta_T}{2}$

$\theta_T = 180° - 2 \cdot \theta$

$D = \dfrac{B}{2 \cdot \tan(P)}$

## Data :

**Activity 1 :** Height Determination of Object on Wall

Calculate Apparent Height Table

| Trial | Top Height | Bottom Height | Apparent Height |
|---|---|---|---|
| | $R_T$ (cm) | $R_B$ (cm) | $\Delta R$ (cm) |
| | | | |
| | | | |

Similar Triangles Data Table

| Trial | Distance from Object | Arm Length | Apparent Height | Calculated Height of Object |
|---|---|---|---|---|
| | $D$ (cm) | $A$ (cm) | $\Delta R$ (cm) | $H$ (cm) |
| | | | | |
| | | | | |

**Activity 2 :** Determination of Height through Angle ( Tangent ) Method.

| Trial | Distance from Object | Angular Height | Calculated Height of Object |
|---|---|---|---|
| | M (cm) | $\Theta$ ($^o$) | X ( cm ) |
| | | | |
| | | | |

**Activity 3 :** Height Determination of Object Outside

Calculate Apparent Height Table

| Trial | Top Height | Bottom Height | Apparent Height |
|---|---|---|---|
| | $R_T$ (cm) | $R_B$ (cm) | $\Delta R$ (cm) |
| | | | |
| | | | |

Similar Triangles Data Table

| Trial | Distance from Object | Arm Length | Apparent Height | Calculated Height of Object |
|---|---|---|---|---|
| | D (cm) | A (cm) | $\Delta R$ (cm) | H (cm) |
| | | | | |
| | | | | |

**Activity 4 :** Determination of Distance of Object through Tangent Method

| Trial | Distance from Object | Angular Height | Calculated Height of Object |
|---|---|---|---|
| | d (cm) | $\Theta$ ($^o$) | X ( cm ) |
| | | | |
| | | | |

**Activity 5 :** Determination of Altitude by the Tangent Method

| Trial | Distance from Object | Angular Height | Calculated Height of Object |
|---|---|---|---|
| | d (cm) | $\Theta$ ($^o$) | X ( cm ) |
| | | | |
| | | | |

**Activity 6 :** Determination of Distance of Star by 'Parallax' Method

| Trial | $\Theta°$ | $\Theta_T°$ | $P°$ | $B(m)$ |
|---|---|---|---|---|
| | | | | |
| | | | | |

## Calculations :

Be sure to use a Slide Rule for your calculations.
It is best to use average values, so do several trials for any given activity in this set.

Formulae to Use :

$$\Theta = 90° - \Theta_2$$

$$\frac{H}{D} = \frac{\Delta R}{A}$$

$$X = d*\tan\Theta$$

$$\Theta_T = 180° - 2*\Theta$$

$$P = \frac{\theta_T}{2}$$

$$D = \frac{B}{2*\tan(P)}$$

For known heights ( such as a picture or poster )
Determine the percent error from your experimental value :

$$\%E = \frac{[\text{Experimental Value-Accepted Value}]}{\text{Accepted Value}}*100\%$$

## Conclusion:

From these activities it is clear that there is a connection between math in the form of algebra and geometry and the real-world when we find the place to use the ideas from math. These techniques are the means to solve the problems since there are many times we cannot verify the answer so readily. Consider the case of a star's distance. The most reliable is the measure of the parallax! All other measuring methods in astronomy have to be calibrated from it.

The key to the activities is patience, proper measurement, doing the work repeatedly and then determining the accuracy of the work when comparing to known values.

Summary :

Notes : From Activity 6, one could argue to use the angles measured from the points along the line and then use the tangent to find the distance. This does work and can be used. I designed the activity this way as parallax for stars is measured by the angle of the star's displacement and not the angle that it is measured with respect to the baseline here at Earth's orbit.


Alternative Outdoor Method :

On a sunny day go outside, measure your height with a measuring tape. With the Sun in place to cast shadows, measure both the height of your shadow and the height of a shadow cast by an object ( such as a flag pole, telephone pole, light pole, tree, et al ).

Using the proportion method, estimate the height of the object from the following :

$$\frac{\text{Your Height}}{\text{Your Shadow's Height}} = \frac{\text{Height of Object}}{\text{Object's Shadow Height}}$$

# Activity #4
## Scale Drawings and using the Slide Rule
Grade Level : Middle School
Math Level : Calculating


Since the Slide Rule is a natural tool for Proportions because the adjacent C & D Scales when placing a value on the C Scale opposite some other value on the D Scale creates a continuous series of similar ratios ( or fractions ) it turns out to be the best tool for the math activity of creating a scale drawing.

Place to Map and Tools to Choose :

The first step is to choose the area to be represented on the map. It may outdoors or indoors. The size of it will determine the best choice of measuring tool. If it is very large in area it may be best to use a trundle wheel measuring tool. This type of area may be large buildings and its surroundings or whole streets or portions of streets with several houses for this. Areas on the scale of the size of a yard can be measured using a long measuring tape. Still smaller areas, such as a room can be done with a regular measuring tape or even a meter stick.

Taking Measurements :
With the choice of area to be measured and the appropriate tools, then take measurements both in a table and with a basic sketch where key objects are measured with reference to various points. Note that all objects need at least two measurements for accurate placement on the maps ( basically an x-direction and a y-direction ). It also helps to have a constant place to take all measurements from since it acts as the focus or anchor of the map. This point can be one of the corner intersections.

Creating a Scale :

Take the longest measurement from your measures ( in meters or feet ) and this is the denominator and is found on the D Scale. In the numerator which is the C Scale is the measurement of the longest measure of the area to be used as the map ( it may the length of the piece of paper used or a box on it – measured in centimeters or inches ).

The ratio on the Slide Rule can be used literally and it is the ratio of the map units to the measured dimensions in the area ( such as inches to feet, centimeters to meters, or even a combination such as centimeters to feet ). For example, let's say the largest distance measured was 75 ft and the paper being used was 24 cm. This ratio, $\frac{24 \text{ cm}}{75 \text{ ft}}$, is then set on the Slide Rule. Read along the D Scale to find a given distance and find its corresponding scale measure opposite it on the C Scale.

Creating the Map :

With the Scale in place now look up each of the dimensional value measurements you have made. You must use the rules of the Slide Rule for decimal placement carefully in these calculations. Realize since we have set the ratio to the maximum, all other map measurements will be less. One other problem you will probably encounter is that measurements are not decimal values. So continuing with our example, let's say one of the measurements was 14 ft 9 inches. We need to look up on the Slide Rule $\frac{9}{12}$ by placing 9 on the C Scale over 12 on the D Scale and reading opposite the D Scale Index the decimal value 0.75. This means our measurement is 14.75 on the D Scale after we reset the Slide Rule for the Scale of the Drawing. In some cases we have to round since the Slide Rule can only obtain 2-3 significant figures. We find for this example on the C Scale a map measurement of : 4.7 cm ( calculator 4.72 cm ).

With all of the measurements converted into the Scale Measures, we place the objects on the map at the required distances to generate the scale version of the real world system we originally measured.

Other Scale Considerations :

We can also fit the map a bit better by choosing a smaller and whole number value Scale. Since this ratio we originally used is the maximum, it is may be best to round it down to a ratio of values of whole numbers that generates a decimal value less than the original ratio. Also it depends on whether metric or English measuring units are being used. If the map units are inches and the real world measurements are feet, then you would want to find ratios such as 1/8, ¼, 3/8, ½, and the like. These types of ratios make the task easier. Say our ratio resulted in a value of 0.32 inches to feet. ¼ is 0.25 while 3/8 is 0.375. Use the smaller value here of 0.25 or ¼. This Scale would be 1 inch to every 4 feet or ¼ inch to each foot of measurement, which with an English measure ruler can readily be found.

Other Scale Activities :

We can do this same Activity even without taking the physical measurements ourselves. Using graph paper where given horizontal lines are considered as lines of latitude ( the scale of the graph paper will determine how many degrees to make each line as well as the scale of the measures we are considering in the real world such as the distance between cities ). The Vertical lines are lines of longitude. We look up the latitude and longitudes of various cities we are considering and place them on the map. We take measurements between the cities using a ruler. The ruler measurement is the C Scale value and the actual distance ( looked up ) is the D Scale value to create our Scale for this map. We can place objects of known latitude and longitude on this map and determine their distances with our Scale. Given distances in the real world can be determined on the map and its scale distance ( though the latitude and longitude may be unknown ) but with three cities and their distance to another place separately, using a compass sweeping out arcs around each city with the span being the distance to the object from that city, where they all cross each other is the location of the object in question.

# Activity #5
## Calculating Geometry Area & Volume
Grade Level : Middle School
Math Level : Calculating


**Purpose :** To explore and calculate the Area or Volume of regular polygons as well as applications of Area and Volume calculations to everyday life and using a Slide Rule in the Process.

## Materials :
- Ruler,
- Measuring Tape,
- Meter ( Yard ) Stick,
- Protractor,
- Items to Measure ( Room, Table, Books, Walls, et al ),
- Slide Rule


## Procedure :

## Making, Measuring, & Calculating Areas and Volumes of Various Geometric Forms :

1) Be sure to know how to use a slide rule ( look to the chapter on directions and/or direct examples in the first Activity
2) The first thing to say is this : For both the Area and Volume Calculations you can draw the figures, write down measurements and then calculate the answers OR you can imagine the figures with their chosen measurements and calculate the answers OR refer to common and agreed upon items with parental approval to take measurements of and do the calculations. In all cases, you are doing the math and should use a Slide Rule. : )
3)
4) **Areas of Geometric Forms :**
5)
6) For this Activity, use Graph Paper, a Ruler, Protractor, pencil, scissors, and your Slide Rule ( of course )
7) A simple exercise is to examine some of the formulae for Area of some of the geometric figures.
8) Start by drawing a Square that is 10 units by 10 units – use the graph paper box divisions for this.
9) Draw a $2^{nd}$ one of these squares.
10) On this $2^{nd}$ square now draw a diagonal line from one corner to the opposite corner so that the square becomes 2 triangles of equal size.
11) Now cut out the second square and cut apart the two equal triangles.
12) By examination, it is easy to see and reach the conclusion that if two triangles of the same size fit into the square that each is one-half of the Area of the Square.
13) Notice the formula for the Area of a Square : $\mathbf{A = s^2 = s*s}$
14)

15) In our case of 10 by 10 the answer is simple and is 100, the exact number of boxes in the square.

16) By simple reasoning it is easy to see that the triangle must be one-half of this value or 50.

17) Look at the Area formula for a Triangle : **A = 0.5\*b\*h** – In this case each of b ( base ) and h ( height ) are 10, so the answer is 50, as expected.

18) This basic exercise can be done for any other regular polygon to test various ideas, such as for a pentagon, where there are 5 equal triangles inside of it.

19) To determine what the Central Angle measure ( All of the angles having a common vertex, as the center of the regular polygon in question ) is for each of these triangles, use the following formula ( n is the number of sides ) :

20)

21) $$\mathbf{CA = \frac{360°}{n}}$$

22)

23) Also since the remaining two angles ( each referred to as OA ) for the triangle in the polygon must be equal, we can use the following formula to find their measures :

24)

25) $$\mathbf{OA = \frac{(\,180°\text{-CA}\,)}{2}}$$

26)

27) The above exercise is to explore the Areas of common regular polygons and relations between them.

28)

29) For further work, we now assume that the Formulae are indeed correct :

30) In this case, use your graph paper, ruler, protractor, and pencil and draw various regular polygons of different sizes, or imagine such figures and then employ the Slide Rule to perform calculations.

31)

**32)      Volumes of Geometric Forms :**

33)

34) In the case of Volume, often it is best to imagine the form being considered. Be sure to look at the formula to see what dimensions are needed. Then make reasonable estimates of what the values would be for a particular item one is considering – look this up or ask for clarification if necessary to validate your estimated values.

35) In any case, try to pick values so that each of the chosen imagined objects can then be compared to each other.

36) A great way to compare any two known Volumes once calculated is to divide one by the other, so then one can see how much larger one is as compared to the other. Look at division rules for this.

37) Some items to consider ( with parental permission and supervision, of course ):
   a.   A book or an end table as a Rectangular Solid ( Notice that one of these is measured in inches or centimeters, while the other is feet, possibly meters )
   b.   Large Rectangular Solids might include a Room or even a House ( excluding the roof )
   c.   A cylindrical flashlight, paper towel roll, or bucket to act as a Cylindrical Solid
   d.   With some ingenuity ( string or a flexible measuring tape ) measure the circumference of a ball ( basketball, baseball, marble ) and determine its volume. Note diameter = 2 times the radius and the Circumference is pi times the diameter ( these are the written formulae to use to find the radius as needed in the volume formula – hint )
38) Another fun thing to do mathematically is to convert one volume into another. The following is a list of possible conversion factors that can be used. It is best to use the Proportion method on the Slide Rule when converting – look to the section here where recipe volumes are examined or the Activity on Ratios & Proportions on the web site.
39)      **Conversion Table List :**
40)      **1 cu. In. = 16.4 cc**
41)      **1 cc = 0.061 cu.in.**
42)      **1 qt = 0.946 L**
43)      **1 L = 1.057 qt**
44)      **1 Gal = 3.78 L**
45)      **1 L = 0.264 Gal**
46)      **1 tbsp = 14.8 mL**
47)      **1 tsp = 4.9 mL**
48)      **1 tbsp = 3 tsp**
49)      **1 cup = 16 tbsp**
50)      **1 tbsp = 0.5 fl. oz.**
51) Note – throughout this Activity there are other lists you might consider making use of as well as the ones noted here.

### Cost per Unit Known ( Mass, Volume, Weight, Item Number )

1) First we will examine Volume of Objects around us that are already known and determine the ratio of cost per unit whatever we wish ( such as unit volume of our choice, unit mass, et al ).
2) Most stores today will provide the cost of an item per unit ounce, per pound, per unit number of the item. The first question then to determine is this : Is it correct?
3) The goal of this activity is to compare it to other brands or the same brand on the shelf only packaged in a different size.
4) Note that often the larger sizes are not labeled in the same manner of cost per unit as the smaller ones.
5) Be sure to read the labels carefully since the unit cost provided might be something like cents per ounce yet the material is labeled in dollars and pounds.

6) Therefore it is a good idea to keep in the back of one's mind basic conversions, such as :

**7) 1 pound = 16 ounces**

**8) 1 cup = 8 ounces**

**9) 1 gallon = 4 quarts**

**10)    1 quart = 2 pints**

**11)    1 pint = 16 ounces**

**12)    1 kilogram = 1000 grams**

**13)    1 kilogram = 2.2 lbs ( pounds )**

**14)    1 ounce = 28.3 grams**

15) With these in mind it is best to choose base units for calculation and comparison, for example turn all units to ounces for mass or volume considerations.

16) Reading a label first determine the total number of ounces if not given.

17) To calculate the cost per unit ounce (CPO):

18) Cost per Unit Ounce (CPO) = $\dfrac{\text{Cost (\$)}}{\text{No. of Ounces}}$

19)  For all other types of costs per unit whatever it is simply the cost divided by the number of ounces ( number of pounds ) (number of items ) and the like.

20) In all cases these are rates and a rate is a ratio or fraction.

21) The best way to treat all ratios is to let the Numerator ( here Cost ) be the C scale and the Numerator be the D scale ( here the number of Ounces ). The answer is read opposite the D scale index on the C scale.

22) Once you know the cost per item it is easy to determine the total cost of a purchase by taking that ratio and multiplying by the number of items you intend to buy.

23) Here we can take the volume of an item, say a gallon of milk and determine the cost per unit volume of another measure, such as the number of quarts in the gallon and compare this to the price per quart in the store.

24) We can also consider the cost of a given cup of milk by first using the ratios above and finding the number of cups in a gallon and determining the cost per cup with the slide rule.

25) Here is a quick example of Cost per unit Ounce :

26) Let's say a can of vegetables in one size of 16 oz has a cost of $1.20, we would find the cost per ounce by placing 1.2 on the C scale over 1.6 ( thought of as 16 ) on the D scale.

27) When it comes to division, the number that is the denominator, here 16 is the scale we read along to find its Index – here the Left Index on D is adjacent to no numbers, but the Right Index is, this is the one we read.

28) We find the answer opposite the Right Index on D on the C scale and read 7.5, but we need to realize the decimal point. Use the Scientific Notation Rules above to find the answer to be 0.075 or 7.5 cents per ounce. ( This is because the exponent in the Numerator is 0, the Denominator is 1 and the slide projected to the right during division so 0-1-1 = -2 our new exponent )

## Determining the Area of Places in the Home :

1) In this next exercise, take the measurements of each of the rooms in your house – the length and width. It is best to create a Table for them. Also it is best to measure in feet and inches.
2) One of the best things to consider doing is to convert the feet and inches into meters and centimeters for those wishing to learn conversion and using the metric system as well.
3) Here is a table of common conversions that may be needed :
4) **1 m = 100 cm**
5) **1 inch = 2.54 cm**
6) **1 ft = 12 inches**
7) **1 yard = 3 ft**
8) **1 sq. ft. = 144 sq. in.**
9) Here is the basic formula needed for all of these calculations :
10) $A = L * W$
11) That is to say, Area equals Length times Width
12) What if you wanted the answer in square feet instead of square inches?
13) The measure of the inches part of each of your measurements needs to be converted into decimal form by dividing it by 12, of course.
14) For example, let's say one of the measurements was 10 ft 8 in.
15) Take the 8 inches and divide by 12.
16) This is 8 on the C Scale of the Slide Rule placed over 12 ( 1.2 ) on the D scale of the slide rule.
17) The Answer is read opposite the Left Index on the D scale and is 0.67
18) For directions of multiplication and division on a slide rule use those listed above or see How to Use a Slide Rule and Using a Slide Rule sections on the Web Site.
19) Now let the other measure be 9 ft 3 in. Use the same directions to convert the inches to feet as a fraction and arrive at 0.25 for it.
20) Now to determine the Area of this room :
21) $A = L * W = 10.67 \times 9.25$
22) First estimate the answer : 10 x 10 is close so the answer is about 100 ft$^2$
23) Since 0.25 is further from 10 than is 0.67, the answer will be less than this, so we can further estimate the answer to between 95 and 100 ft$^2$
24) Place 10.67 ( note the estimation of the last digit ) on C scale over the Left Index of the D scale.
25) Now read along to 9.25 on the D scale and read the answer above on the C scale.
26) Looking carefully you should note that the .25 mark is just to the right of 98.5 so we would read this as 98.6 for 98.6 ft$^2$.
27) ( A check on a calculator yields 98.6975, but of course nobody measures a room to 4 decimal places – 1 is sufficient, which the slide rule provides accurately ) –
28) In fact from our measurements and calculation we can only have 3 significant figures at best in the answer and coming to 98.6 when the calculator reads 98.7 is quite good!

29) A far better realization is this : What if you were to carpet or tile this room. You would not use the 98.6 value anyways. It is better to overestimate and buy more than is needed, otherwise there may be problems. In this case even 100 ft² is too close and it would be better to round each measurement up to the next whole number of 11 and 10 and estimate the area at 110ft².

## Home Project Needs for Carpeting and its Cost

1. The most basic of calculation is simple enough, first measure the Length ( L ) and Width ( W ) of the room in the chosen units ( ft or m ) and calculate the Area ( A ). ( A = L * W )
2. To make the process simpler and avoid mistakes round all values up to the next whole number value.
3. *Not Recommended But For those who like some level of precision, in the case of inches divide the inches past the feet measure by 12 and tack this decimal value onto your feet measure, instead of rounding up.*
4. Determine the Number of Square Yards Needed by taking the Total Area ( in square feet so far ) and divide by 9.

   a. **Number of Yards Needed** $= \dfrac{\textbf{Total Area in sq ft}}{\textbf{9}}$

5. Note that this Calculation is for Vinyl Flooring too! Follow the same procedure noted above down to the cost here below.
6. The final calculation is cost, simply take the Cost per yard ( CY ) on C scale place over 1 on the D scale and read along the D scale to the needed number of yards needed ( N ). The Total Cost ( TC ) is found on the C scale above. TC = CY * N

   a. **Home Project Needs for Tiling a Floor and its Cost**

7. The most basic of calculation is simple enough, first measure the Length ( L ) and Width ( W ) of the room in the chosen units ( ft or m ) and calculate the Area ( A ) as we did above. ( A = L * W )
8. To make the process simpler and avoid mistakes round all values up to the next whole number value.
9. *Not Recommended But For those who like some level of precision, in the case of inches divide the inches past the feet measure by 12 and tack this decimal value onto your feet measure, instead of rounding up.*
10. Determine the Number of Tiles Needed by taking the Total Area ( in square feet so far ) and divide by The Area in Square Feet of 1 Tile ( Note : One Tile may be 1 piece, such as 3" x 3", et al, but can also be the square of small tiles on a mesh that is places down – these can be 6" x 6", 9" x 9", and other sizes ).

11. First determine the Area in Square Feet of 1 Tile : Once again, Area is Length times Width, but typically measured in inches, so your answer is in square inches. To convert to Square Feet divide this answer by 144. Now Take this number and use it in the following formula to determine the Number of Tiles needed :

    a. **Number of Tiles Needed** $= \dfrac{\text{Total Area in sq ft}}{\text{Area in Square Feet of 1 Tile}}$

12. The final calculation is cost, simply take the Cost per yard ( CY ) on C scale place over 1 on the D scale and read along the D scale to the needed number of yards needed ( N ). The Total Cost ( TC ) is found on the C scale above. TC = CY * N


## Home Project Needs for Painting and its Cost

1. The most basic of calculation is simple enough, first measure the Length ( L ) and Height ( H ) of a wall in a room in the chosen units ( ft or m ) and calculate the Area ( A ). ( A = L * H )
2. To make the process simpler and avoid mistakes round all values up to the next whole number value. Most walls are 8 ft tall for example.
3. Be sure to sum up all of the walls. TA = total area ( TA = Sum of all Areas )
4. Not Recommended , But For those who like some level of precision, in the case of inches divide the inches past the feet measure by 12 and tack this decimal value onto your feet measure, instead of the rounded up values used.
5. For still greater precision, you can subtract out non-painted areas, such as windows and doors ( A = L*W ) if you like.
6. Whatever the case, take the final number TA and divide by 300. This gives the number of gallons needed for rough, textured, or unpainted wallboard.
7. If it is smooth walls instead divide TA by 350.
8. Note that these are estimates. Always overestimate and round up. For example you are not going to buy 3.3 gallons, buy 4.
9. The final calculation is cost, simply take the Cost per gallon on C scale place over 1 on the D scale and read along the D scale to the needed number of gallons. The Cost is found on the C scale above.


## Home Project Needs for Wallpapering and its Cost

1. The most basic of calculation is simple enough, first measure the Length ( L ) and Height ( H ) of a wall in a room in the chosen units ( ft or m ) and calculate the Area ( A ). ( A = L * H )
2. To make the process simpler and avoid mistakes round all values up to the next whole number value. Most walls are 8 ft tall for example.
3. Be sure to sum up all of the walls. TA = total area ( TA = Sum of all Areas ) This is the Wallpapering Area.

4. For still greater precision, you can subtract out non-papered areas, such as windows and doors ( A = L*W ) if you like, instead of rounding up. Subtract the total of the non-papered areas from the papered ones
5. Use the chart below to find the Usable Yield Value. This is divided into the Wallpapering Area to determine the Number of Single Rolls Needed.

**a. No. of Single Rolls $= \dfrac{\textbf{Wallpapering Area}}{\textbf{Usable Yield}}$**

6. The final calculation is cost, simply take the Cost per roll on C scale place over 1 on the D scale and read along the D scale to the needed number of rolls. The Cost is found on the C scale above.

| Pattern Repeat ( Drop ) | Usable Yield ( American rolls ) | Usable Yield ( European rolls ) |
|---|---|---|
| 0 to 6 in. | 32 sq ft. | 25 sq ft. |
| 7 to 12 in. | 30 sq ft. | 22 sq ft. |
| 13 to 18 in. | 27 sq ft. | 20 sq ft. |
| 19 to 23 in. | 25 sq ft. | 18 sq ft. |

## Determining Recipe Needs through Proportions

1. These directions addresses the question of what to do when a recipe does not match your materials on hand and/or the question of how much is needed when you are changing the scale of the recipe to a larger or smaller yield.
2. The reason it is here in this Activity is because it concerns volume. This application is changing the volume of materials needed in the recipe.
3. This is the type of activity that the slide rule does very well since it is a proportion, which slide rules excel at.
4. The idea to solve this comes from noting that the ratio of the needed amount for a given ingredient in the recipe of a given number of servings will equal
5. The amount you need to have for your batch divided by the number of servings you wish to make!

$$\frac{\textbf{Recipe Requirement for Material}}{\textbf{Recipe Number of Servings}} = \frac{\textbf{The amount of Material Needed}}{\textbf{Number of Servings}}$$

6. This type of logic can be used for any and all variations of the same theme here. Remember to let the Numerator be the C Scale while the Denominator is the D Scale.
7. First place the known values of the Recipe over each other and look along the D Scale to the Number of Servings you wish to make and find the amount of the unknown needed material above. Be sure to employ the proper use of scientific notation when needed.
8. Keep in mind, you can determine how many servings you are making by looking in the opposite direction and going from what you have and looking back on the D Scale at how many servings it will make!

## Data :

Note : The only data needed is your measurements, so create tables, labeled diagrams, and the like as needed from the Activities you engage in.

## Calculations :

Be sure to use your Slide Rule !

## Area, Volume Formulae :

| OBJECT | PERIMETER | AREA |
|--------|-----------|------|
| Square | =4*s | =$s^2$ |
| Rectangle | =2*L + 2*W | =L*W |
| Triangle | =A+B+C | =$\frac{1}{2}$*b*h |
| Circle | =$\pi$*d<br>=2*$\pi$*r | =$\frac{1}{4}$*$\pi$*$d^2$<br>= $\pi$ *$r^2$ |

Note :
s = side;
L = Length, W = Width;
A,B,C are lengths of sides, b = base, h = height;
d = diameter, r = radius

## Volume :

**Cube : V = $s^3$**
( s = side )

**Rectangular Solid : V = L*W*H**
( L=length, W=width, H=height )

**Cylinder : V = $\pi$*$r^2$*h**
( r = radius, h = height )

**Sphere : V = $\frac{4}{3}$*$\pi$*$r^3$**
( r = radius )

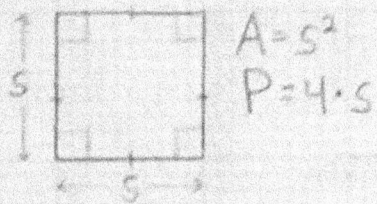

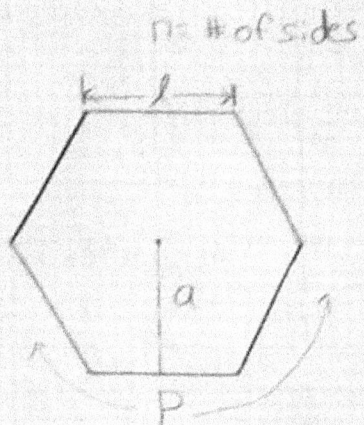

$A = s^2$
$P = 4 \cdot s$

$n = \# \text{ of sides}$

$A = \frac{1}{2} \cdot a \cdot P$

$P = n \cdot \ell$

(for all Regular Polygons)

$A = b \cdot h$

$A = \frac{1}{2} \cdot b \cdot h$

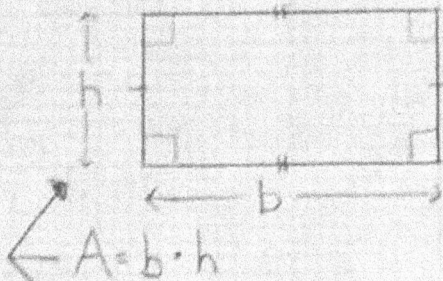

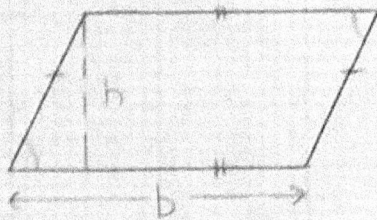

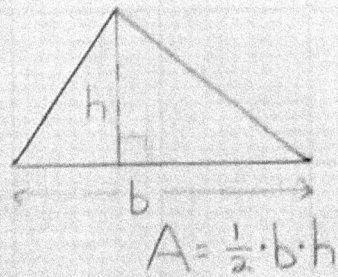

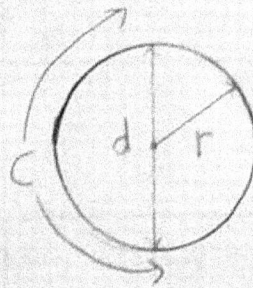

$C = \pi \cdot d$
$C = 2 \cdot \pi \cdot r$
$d = 2 \cdot r$
$A = \pi \cdot r^2$

## Conclusion :

Hope this Activity helps you find some fun and applications to Area and Volume, the use of Formulae, and practice using your Slide Rule.

There are things you can extend the idea of volume to, such as found in other Activities, such as Cost of Water Usage as well.

There are a number of other Geometry Activities as well on the web site, such as Determining Pi in two different Activities in two different ways, Using Right Triangles and finding the unknown hypotenuse length, treating vectors at right angles to each other as a right triangle to find the determinant, et al. have fun : )

## Activity #6
## Finding Pi through Geometry
Grade Level : Middle School
Math Level : Calculating

## Archimedes Geometric Method of Determining Pi ( $\pi$ ) Activity

It is hard enough to draw a perfect circle, let alone measure it accurately to determine its circumference. This was the problem faced by Archimedes. It is best to find a way to analyze this curved object by means of straight lines, he came to reason.

Archimedes, is considered one of the greatest of mathematicians and inventors of all time. Archimedes of Syracuse ( c. 287 BC – c 212 BC ) is best described as a Greek mathematician, inventor, engineer, physicist, and astronomer. You can find more about him in various Activities with his name mentioned in the title.

He gave us a very useful range and value for ☐pi of 3.14 used even to this day. His work found it to be a value between $3\frac{1}{7}$ ( approx. 3.1429 ) and $3\frac{10}{71}$ ( approx. 3.1408 ) by using his method of determining the perimeter of a regular polygon around a circle and one inscribed on the inside.

What if you could just inscribe a circle inside a regular polygon, such as a square, a pentagon, a hexagon, or some other multi-sided regular polygon ( note : a regular polygon has equal length sides ). In this case the circle should only touch the polygon at one point on each section of the polygon. For example, if the regular polygon were an octagon, then it would touch the circle on eight points. Measuring this polygon's perimeter is as easy as measuring the length of the sides and summing them up. Easier still considering the fact that each side length is the same, so knowing the length of one side, all one has to do is multiply by the number of sides to find the perimeter. Being outside the circle, it is clearly longer than the circle's perimeter (circumference), so this would be an upper bound of that value.

Now inscribe a regular polygon inside the circle in exactly the same manner so that it touches on one point per side of the regular polygon. The perimeter of this inscribed regular polygon will provide then a lower limit to the circumference of the circle. We now have an upper and lower limit to the actual circumference of the circle. ( One method other than the picture above is to use the same type of polygon you used in the other case, only here rotate it an angle ½ of the central angle value ).

If we did this process and made a table and kept increasing the number of sides, at an infinite number of sides, it would be the circle itself. Hence the larger the number of sides the closer to the actual values it would come.

112

Better still, we can take these upper and lower circumference values and divide by the diameter of the circle in question that is nestled between these regular polygons.

Today we know this ratio as Pi ( $\pi$ ) = $\frac{\text{C-circumference}}{\text{d-diameter}}$ . We know its value to be approximately 3.14*** In fact this is what Archimedes did when doing this type of process in a similar manner and arrived at a value we use for pi ( $\pi$ ) and was used for many centuries.

In our Activity we will draw one set of inscribed and exterior regular polygons about a circle and do this very process. It is best to start with the least number of sides and continue upwards in number – such as 8, then 12, then 18. You can pick any number you want, but be sure to do the calculations correctly since each has a different central angle value. ( As an aside here are some names : 4-sided : quadrilateral ( officially ), 5-sided : pentagon, 6-sided : hexagon, 7-sided : heptagon, 8-sided : octagon, 9-sided : enneagon, 10-sided : decagon, 12-sided : dodecagon, 15-sided : pentakaidecagon, 16-sided : hexakaidecagn, 18-sided : octakaidecagon, 20-sided : icosagon, 24-sided : icosikaitetragon, 30-sided : triacontagon, 50-sided : pentacontagon, 100-sided : hectogon ). We will use our polygons to estimate the circumference of the circle, to which we can compare and also use this value to find an upper and lower boundary to pi ( $\pi$ ) and compare this value to pi ( $\pi$ ) itself.

**Purpose :** To use two regular polygons, one exterior and one interior to a given circle to determine the upper and lower limits to the value of the ratio of circumference to diameter or pi ( $\pi$ ) in a manner parallel to Archimedes.

## Materials :

- Graph Paper,
- Protractor,
- Ruler ( best to have both English and Metric measures on it ),
- Compass,
- Spanner ( possibly – as found in a compass set ),
- Slide rule

## Procedure :

1. Note : Measurements of the polygon sides can be done in any of 3 ways :
2. The first 2 ways are simply either English or Metric measurements. It is recommended to do both in this exercise.
3. The Metric measurements allow for decimal use and estimating to the nearest $1/10^{th}$ of a millimeter.
4. The English measurements allow for fraction use and computations involving them.

5. The final method is to either use the provided formulae in the Calculations section or you can derive them from looking at the process of each of the polygons you create and the formulae that you arrive at from the process of determining the length of the side of the polygon and then finding the perimeter.

6. The following steps can be used for each of the chosen polygons you have decided upon constructing where there is an external polygon and an internal polygon to the circle you make.

7. It is recommended to use the following polygons in this order : the 8-sided Octagon, the 12-sided dodecagon, and then finally the 18-sided octakaidecagon.

8. Steps in the Procedure are as follows :

9. It is best to start in the middle of the graph paper ( and best to use large square variety – about 4 squares to the inch ).

10. Draw a conventional axis system ( not labeled ) along the graph lines. The central point will act as the center of both the circle and the polygons.

11. Choose a distance from the center that will act as the radius of the circle ( r ).

12. Note : It is best to choose a whole number of squares ( if you want ), but any number will do. Also note that you can use either metric or English metric measuring for this measure ( or even just a number of squares at this moment, which will later be measured ).

13. Along each of the axes place a dot at the chosen distance.

14. Determine the central angle measure ( $\Theta$ ). Record this result.

15. Place the protractor on the axes and measure off the determined central angle value in sequence all the way around the circle. Place a mark at each of these points. ( For example, a hexagon would have a central angle measure of 60°, so there is a mark each 60° ).

16. Connect the points through the center of the circle. Use one type of writing tool ( for example pencil ).

17. Measure out along each ray from the center the prescribed distance of the radius ( r ).

18. Connect these points with a ruler. Use the same writing tool for the lines you have made.

19. Now create the next polygon :

20. Divide the central angle by 2 and starting on the positive x-axis, use the protractor to measure this new angle and place a mark there ( in essence it bisects the angle made with the x-axis and the first measured central angle ray ).

21. From this new 'starting point', now mark off with a new writing tool ( another pencil of a different color or a pen ) the angles with the same measure of the central angle measure. ( In essence, each of the prior marks are being bisected ).

22. Like the prior polygon, measure out along these new rays the distance of the radius of the circle ( r ) and make a mark with the other writing tool.

23. Connect the points with the other writing tool.

24. At this point, if done correctly, there are two regular polygons with the same number of sides and one is rotated in a manner to be half of the central angle for each section.

25. Create the circle : ( Note this could be done first if you want ) :

26. Using the compass open it to the span of the radius and create a circle with the center as the center of your axes system. If done correctly it will only touch one point in each section of your inscribed and exterior regular polygons.

27. To make measurements, it is best to use the compass or a spanner ( oftentimes found with compass sets ) to span the length of the sides.
28. First decide to do the exterior or the interior polygon.
29. In each case, choose to do a set number or all of the sides, whereby you span the length of the side with the compass or spanner and then place the points on the ruler on the chosen measuring units ( metric, English ) and measure it as noted above. Record these results.
30. Use your results for a given polygon ( exterior or interior ) side lengths and find the average value for each separately.
31. Sum up each of the polygon sides for the perimeter of each of them ( the exterior and the interior ). The exterior should be larger and when divided by the diameter of the circle, should have a value greater than pi ( $\pi$ ) while the interior perimeter value divided by the diameter should be less than pi ( $\pi$ ).
32. Continue this process for each of the polygon sets you decide to do in determining pi ( $\pi$ ). Compare results from one trial to the next. What happens as you increase the number of sides ( are your values closer or further from pi ( $\pi$ ))?
33. Take the sum of these two determined values of pi ( $\pi$ ) and take their average and see how close this comes to pi ( $\pi$ ) itself.
34. For those interested in the geometry of the situation :
35. Look at each of the isosceles triangles of both the interior and exterior polygons and know that you know the angles and the length of the radius ( r ) – when the triangle is bisected it is two right triangles, and it may be easy to find half of the side length through using trigonometric relations. ( Hint : be careful as to which side of the triangle is 'r' – in one case it is a hypotenuse and the other it is not ). – you need to determine the corner angles in the triangles ( $\Theta c$ )
36. If you have done the prior step, how close do your measurements for the side of the polygons ( x ) come to the mathematically determined values? – For help look to the provided formulae if needed, or check your answers!
37. Using these mathematically determined values of the sides of the regular polygons now determine the perimeters and the values of pi ( $\pi$ ) as you did before to see how close your values come to it.
38. Side Note : Did you see that using metric measurements allows you to use decimals, examine precision and accuracy? Or in the case of the English measuring system, not only is there precision and accuracy, but here there is the option of using fractions?

## Diagrams for Example :

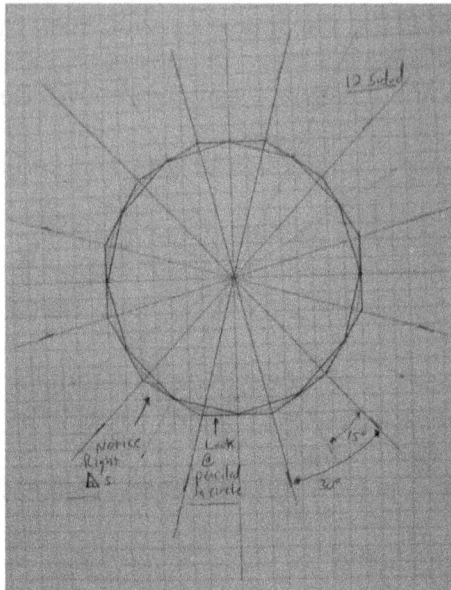

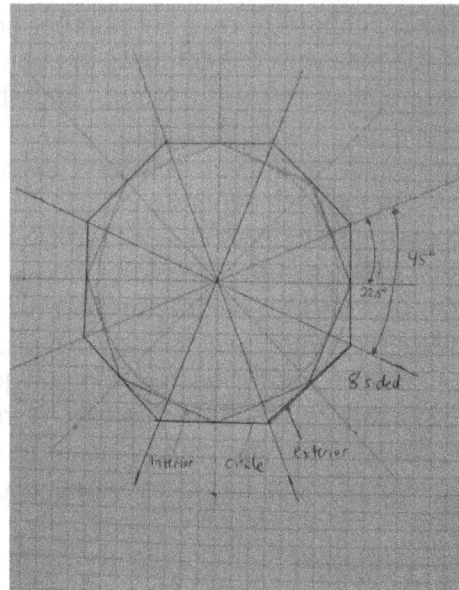

## Data :

For a chosen Polygon ( be sure to do 3 different ones ):

Number of Sides ( n ) : _____
Chosen circle radius ( r ) : _____ ( unit of measure – metric, English )
Central Angle ( $\Theta$ ) : _____ ( $^{\circ}$ )
Corner Angle ( $\Theta c$ ) : _____( $^{\circ}$ )

Measures of Sides : ( x )

| Side | Measure ( unit ) |
|------|------------------|
| 1    |                  |
| 2    |                  |
| ...  |                  |
| N    |                  |

## Calculations :

Be sure to use your Slide Rule!

## Central Angle Formula :

$$\Theta = \frac{360°}{n}$$

## Isosceles Triangle Formula :

$$\Theta + 2\Theta_c = 180°$$

$\Theta c$ is the value of the two congruent angles

$$Sin(\Theta) = \frac{length\ of\ side\ opposite}{length\ of\ hypotenuse}$$

$$Tan(\Theta) = \frac{length\ of\ side\ opposite}{length\ of\ side\ adjacent}$$

Note : Though □is used here, it can be any '$\Theta$' as needed! ( hint )

Diameter and Radius Relation :

$$D = 2*r$$
$$D = diameter,\ r = radius$$

## Circle Relation :

$$\pi = \frac{C\text{-Circumference}}{D\text{-Diameter}}$$

Derived Side Distance Formulae :

Note : Do not look at these if you want to derive them
Note : x = side length, r is a radial measure length to the circle,
n is the number of sides for the polygon

Interior Polygon Side Length ( x ) :

$$x = 2*n*r*\cos(\Theta_c)$$

Exterior Polygon Side Length ( x ) :

$$x = 2*n*r*\tan(\Theta_c)$$

Average :

$$x = \frac{\Sigma x}{n}$$

Value of Pi ( $\pi$ ) :

$$\pi = 3.1415***$$

## Conclusion :

How close did your values for your outer and inner polygons come to the
value of pi ( $\pi$ )? As you increased the number of sides, what did the
values for the upper and lower limits in your calculation do in relation to
pi ( $\pi$ ) – did they come closer or where they further away?
Compute a percent error to the appropriate number of significant figures
for your figure – To do this : subtract your determined value from the
accepted value and divide this difference by the accepted value. The
result is a decimal representation of your error. Multiply by 100 for
the percent.
Could you generalize the formula so that you can keep increasing the
number of sides and then computationally determine the upper and lower
limits in approaching pi ( $\pi$ )?
When using the slide rule, did you notice a pattern for the upper and
lower limits as to where it fell on either the C or D scale in relation to the
$\pi$ symbol?

# Activity #7
## Vector Addition and the Slide Rule
Grade Level : High School
Math Level : Challenging

In Science, there are two basic types of quantities, Scalars and Vectors. Scalars are those that have only magnitude ( amount ) and include such things as Mass, Temperature, and Time. Vectors are actually the majority of Sciences, such as Physics. Vectors are quantities that have both Magnitude ( Amount ) and Direction ( often denoted as an Angle and measured from a reference point ). Some of the more common vectors are : Velocity, Acceleration, and Force. This includes all forces, such as electric, magnetic, elastic, or any other mechanical force.

Just from the description it is easy to see that vectors play a major role in characterizing and understanding nature. The more complete the understanding of the types of vector quantities affecting any given object under consideration allows for a better model of how nature operates.

If we add two scalar amounts, for example two masses, we merely add them, since their total is the sum of the masses in question. This type of math is not the same for vectors because they are not only amounts, but also these amounts are pointed or moving in a given direction.

There are some obvious answers to vector addition, however. If two vectors are in the same direction they are merely added to give a sum or total amount. This, incidentally, is the maximum resultant vector for these two vectors when being added. To find the minimum value, merely have the vectors point in opposite direction and so we would only subtract them to find their resultant value.

If the vectors acted at right angles to each other on the same object, for example imagine a stream flowing from west to east and a boat crossing it using a constant rate motor and headed from south to north. Each of these velocity vectors is perpendicular and hence we can use the Pythagorean Theorem ( $A^2 + B^2 = C^2$ ) to find the resultant vector magnitude ( here it is C ). Note that this value must be between the maximum and minimum values noted earlier, but not necessarily the average of the two values!

For example, what if the two vectors are 3 and 4 respectively. If both in the same direction, the resultant vector is 7. If in opposite directions, the answer is 1. If at right angles to each other, the resultant vector is 5.

The next logical question is, then, what if the vectors are not parallel nor perpendicular to each other, but instead at some other angle. There must be a system to find the answer to their addition.

The most commonly practiced method is called the Vector Component Method. It takes each of the vectors in question and breaks them into convenient x- & y- components, which are then summed up. Each of these is squared and summed again. Then the square-root is taken. This final answer yields the resultant vector magnitude. But recall that vectors are both magnitude and direction.

The direction is found also in the resultant components. With them the tangent of the resultant angle for the given case can be determined.

The following Activity has the steps necessary to learn such a process, the tools to help this process ( the Table ) and several other problems for you to try for yourself for practice. Enjoy!

**Purpose :** To use a Slide Rule to Add 2 Non-Parallel, Non-Perpendicular Vectors
to generate a Resultant Vector.

## Materials :

1) Table or Story Problems of Vectors to be added,
2) Slide Rule

## Examples & Procedure :

1) In Vector Addition one of the best methods ( and most commonly used ) is the Vector Component Method for Vectors that are Non-Parallel ( here, if parallel, we would merely add or subtract them as needed ) and Non-Perpendicular ( here we would simple use the Pythagorean Theorem, if perpendicular ). Here, the Vector Component Method will be used.
2) Note that other techniques are available, but not explored here. ( The most common is the expanded Pythagorean Theorem ).
3) In the case of all Vectors, we need to have both the Magnitude ( size or amount ) of the Vector and the Direction that this Vector is acting in (oftentimes designated by an Angle).
4) Note : Read below in the Template for directions and considerations when the angle is NOT measured from the +x-axis. For the examples and problems given here, all are measured from the +x-axis.
5) In the example there is a $3.0 \times 10^3$ N force acting at angle of $24°$ to the +x-axis on a given object along with a $4.0 \times 10^3$ N force acting at angle of $48°$ to the +x-axis.
6) Note that only in the case of these two vectors being at right angles to each other would the magnitude of the resultant vector be $5.0 \times 10^3$ N, **BUT** this is **NOT** the case here. When it comes to vectors, read the numbers carefully and given time, check your answers!
7) Also, the process of Vector Addition is really a process of taking many small steps.
8) Each of the Vectors is placed in the Table Template noted below and each of the Components ( x & y ) are determined, summed up and then the needed formula to find the resultant magnitude and angle are employed.
9) For example, let's start with $F_{1y}$. $F_{1y} = F_1 * \sin\Theta_1$.
10) Here $F_{1y} = (3.0 \times 10^3) * \sin(24°)$.
11) First look up on the S Scale ( Sine ) $24°$. Place the cursor here and the answer is found on the D Scale. It reads 0.406.
12) You do not need to write this down nor remember it. Merely leave the cursor here and now slide the Slide so that the Left Index of C is over this value on the D Scale.

13) This is because our goal is to multiply this number by $3.0 \times 10^3$.
14) Now move the cursor to 3 on the C Scale and find the product on the D Scale below. It reads 1.22 which we write as $1.22 \times 10^3$ or 1220.
15) This answer is the y-component of $F_1$.
16) We follow a similar process for the x-component for F1, only there is a problem!
17) There is no 'Cosine Scale' on the Slide Rule. Or is there?
18) Recall that $Cos(\Theta) = Sin(90°-\Theta)$.
19) So here $Cos(24°) = Sin(90°-24°) = Sin(66°)$
20) We can now redo the process using this Complimentary Angle.
21) This yields an answer of $F_{1x} = 2740$
22) Go ahead and do the second vector on your own. It is the same process.
23) The answers are : $F_{2x} = 2680$ and $F_{2Y} = 2970$
24) $F_{RX} = 2740 + 2680 = 5420$
25) $F_{RY} = 1220 + 2970 = 4190$
26) $F_R{}^2 = (F_{RX})^2 + (F_{RY})^2$
27) You have to look up the squares of each of the values noted above ( $F_{RX}$ & $F_{RY}$ ) and then sum these up $2.94 \times 10^7 + 1.76 \times 10^7 = 4.7 \times 10^7$
28) This has an even number of digits so we use the right hand side of the A Scale and find the square-root on the D Scale.
29) Here we find it to be $6.85 \times 10^3$ for the magnitude of the vector.
30) Note that if our values were 3.0 and 4.0, then our answer should read $6.9 \times 10^3$. Notice that the Slide Rule does a sufficient job at providing an answer to our work.
31) The direction is found by taking the inverse tangent of the ratio of the y-component over the x-component.
32) $Tan(\Theta_R) = \dfrac{F_{RY}}{F_{RX}} = \dfrac{4190}{5420}$
33) Find this ratio on the C Scale over the D Scale. The answer is opposite the D Scale Index. Read further up to the Tangent, T Scale and find the angle as 37.8°.
34) Here, then, is the final answer :
35) The Resultant Vector of the addition of the two vectors in question is : $6.85 \times 10^3$ N at angle of 37.8° to the +x-axis. We have even provided more decimals than are required from the problem. We could reduce it to two significant figures instead.
36) The next main question is this : What if there is an angle for a given vector that is not in the first quadrant?
37) The most important aspect of this is one must be certain as to which Quadrant the vector is in.
38) If measured from the +x-axis use the provided list in Calculations to determine which Quadrant it is in :
39) Let's consider an angle such as 110°.
40) From the List, It is clearly in the Second Quadrant.
41) This means that x-components are negative and y-components are positive. ( see Table in Calculations below for this information )
42) How do we find this angle on the Slide Rule?
43) $180° - 110° = 70°$. This supplementary angle will have the same absolute magnitude values for sine and cosine as 110°.
44) Notice that this is the absolute magnitude.

45) Since the x-component is negative, its cosine value must be the negative part of the equation. So Cos(110°) = -Cos(70°).
46) In all cases, measure back to the x-axis to find the needed angle for a given consideration.
47) With each, take your time, know which step you are on and what the goal of each part is. Always check your answers. Enjoy!


## Vector Template, Photos & Notes :

- The following Template is for each set of two vectors. The vectors in question are 1 & 2 and listed in the first column. The x-components for each of them are in the middle column and in the last column are the y-components for each of the vectors.
- As long as the angle is measured from the +x-axis, these formulae are the ones to use.
- If you were to encounter a measure from the +y-axis, there are two ways to consider this :
  - Simply change the sine and cosine parts of the formulae, for example : $F_{x1} = F_1*\sin\Theta_y$ – OR –
  - Change the angle ⬚ into one as measured from the +x-axis. If East of the +y-axis, use 90°-$\Theta_y$. If West of the +y-axis, use 90°+$\Theta_y$.
- Note : Other types of problems may use a Map orientation of the vectors. If so, North is the +y-axis, East is the +x-axis.
- Note that with the Table and the Vectors given they are denoted as 'F', which can imply Forces. However, this process applies to any and all vectors, such as Forces of any type and includes others like Velocity.
- Force questions can involve the same type of forces acting on an object, such as two ropes attached to a crate at different angles. They can also be different forces, such as force due to gravity, force due to friction, electrostatic forces, magnetic forces, force due to wind pressure, and the like. A given object may have two ( or more ) of these acting on it at various angles.
- Velocity Vector Addition Problems often involve a plane with winds from some given direction or a motor-powered boat crossing a stream at some angle and the stream flowing across the bow of the ship.

|  | $F_X$ | $F_Y$ |
|---|---|---|
| $F_1$ | $F_{x1} = F_1*\cos(\Theta_1)$ | $F_{y1} = F_1*\sin(\Theta_1)$ |
| $F_2$ | $F_{x2} = F_2*\cos(\Theta_2)$ | $F_{y2} = F_2*\sin(\Theta_2)$ |
|  | $F_X =$ | $F_Y =$ |

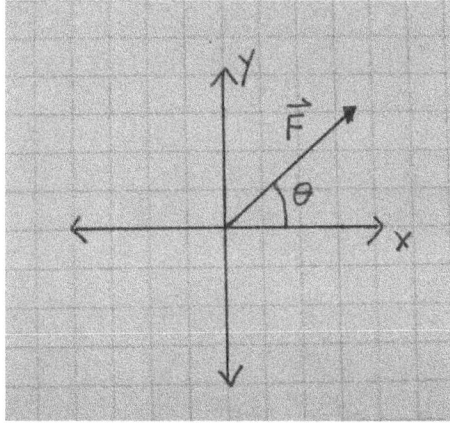

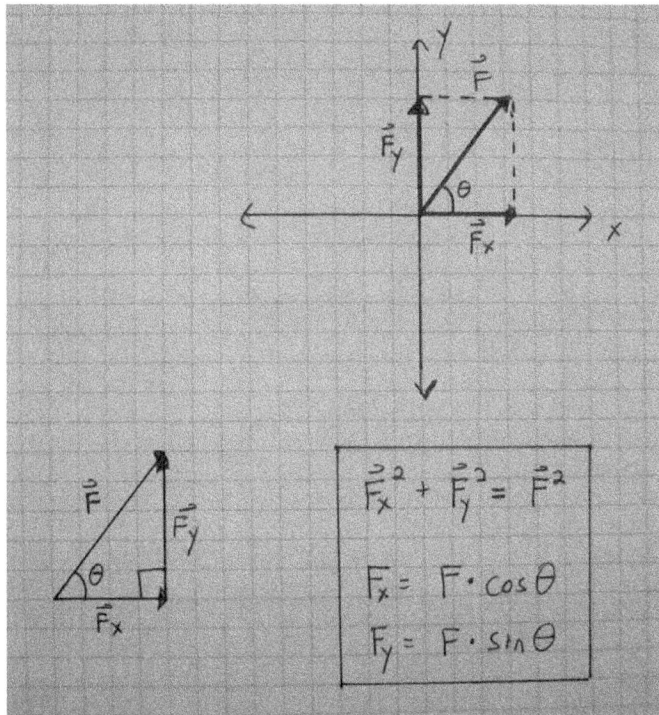

$$\vec{F}_x{}^2 + \vec{F}_y{}^2 = \vec{F}^2$$

$$F_x = F \cdot \cos\theta$$

$$F_y = F \cdot \sin\theta$$

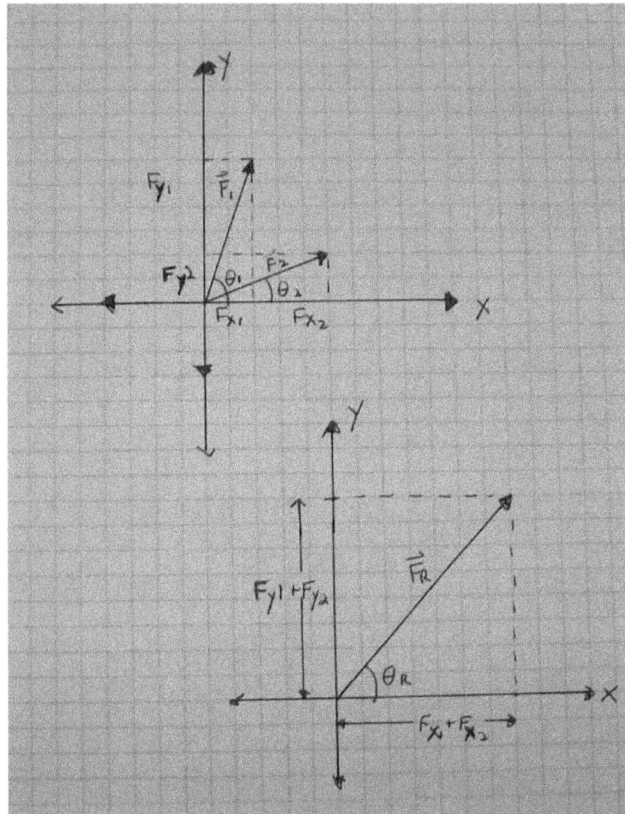

## Vector Addition Questions :

Note the Vectors are listed in order of their Magnitude and then the Angle as measured from the +x-axis in the Table below :

| Question | $F_1$ | $\Theta_1$ | $F_2$ | $\Theta_2$ |
|----------|-------|------------|-------|------------|
| 1 | 78 N | 30° | 45 N | 38° |
| 2 | $4.5 \times 10^4$ N | 45° | $5.2 \times 10^4$ N | 120° |
| 3 | 265 N | 72° | 325 N | 152° |
| 4 | 0.19 N | 50° | 0.18 N | 206° |
| 5 | $8.4 \times 10^4$ N | 115° | $6.2 \times 10^4$ N | 215° |

124

## Calculations :

Be sure to use your Slide Rule!

## x-component formula :

$$F_{XN} = F_N * \cos(\Theta_N)$$

## y-component formula :

$$F_{YN} = F_N * \sin(\Theta_N)$$

## Resultant Vector Magnitude Formula :

$$F_R{}^2 = (\quad)^2 + (\quad)^2$$

Note : This is essentially the Pythagorean Formula!

## Resultant Vector Angle Formula :

$$\tan(\Theta_R) = \frac{\Sigma F_Y}{\Sigma F_X}$$

## Changing Angles as Needed :

$$\cos(\Theta) = \sin(90° - \Theta)$$

## Supplementary Angles :

$$Angle(A) + Angle(B) = 180°$$

**Quadrants :**
**If measured from the +x-axis :**
**All Angles 0° − 90° are in the First Quadrant,**
       **(both x & y components : + positive)**
**All Angles 90° − 180° are in the Second Quadrant,**
       **(x-components : −negative, y-components : +positive)**
**All Angles 180° − 270° are in the Third Quadrant,**
       **(both x & y components : negative)**
**All Angles 270° − 360° are in the Fourth Quadrant**
       **(y-components : −negative, x-components : +positive)**

## Conclusion :

Vector Addition is a classic combination of many aspects of math
( algebra, geometry, trigonometry ) and strongly connects to science in
the real world ( since often things can be considered to act in more than
one dimension and have more than one force acting on it ).

Here is the Answer Key to the Problems :

| Question | Resultant Vector | Resultant Angle ( from +x-axis ) |
|---|---|---|
| 1 | 122.7 | 33.0° |
| 2 | $7.70 \times 10^4$ | 85.7° |
| 3 | 367 | 124° ( from -56° ) |
| 4 | $7.71 \times 10^{-2}$ | 121° ( from -59° ) |
| 5 | $9.53 \times 10^4$ | 205° ( from +25° ) |

# Activity #8
## Using the Slide Rule to Determine Powers ( and Roots )
Grade Level : High School
Math Level : Calculating

The Slide Rule and determining Powers

➢ Though the title mentions roots, this Activity is an illustration through examples on how to use the Slide Rule and the remarkable math power to be found in the L scale when it comes to raising any given number to any given power quite easily.
➢ With regards to roots, the ideas presented here are the same, only instead of multiplying by a power greater than one, roots are values between zero and one, hence can be thought of as fractions which too can be multiplied by the base.
➢ To illustrate the power of Logs, we will try for this activity to use the L scale which takes the log of the number that is looked up on it.
➢ All of these problems are $X^N$ , where X is the base and N is the power. In this example we are not using the A or K scales since our problem will be more complicated than that.
➢ Our problem is this : $(3.4)^{4.1}$
➢ We look from D scale at 3.4, the base, and read the log of the base from the corresponding values on the L scale below.
➢ Looking carefully, the cursor is just to the right of 0.53, and lines up with .532
➢

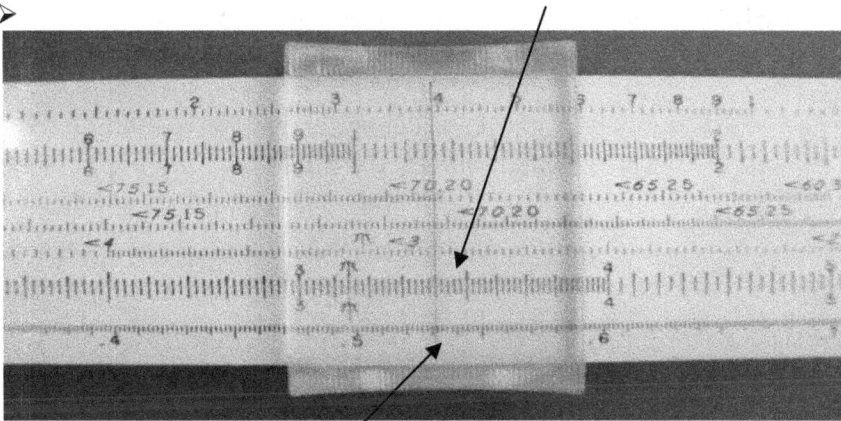

➢ Move the Right Index of the C scale to 532 on the D scale. Note the estimated placement for the 2 in this case.
➢ We are using the Right Index because the value of 4.1 will allow us to stay on the scale and here we are using the rules for exponents which state when taking the log of a value to a given power, the answer is the exponent times the log of the value in question.
➢ Read along the C scale to 4.1 and look below on the D scale.
➢ The D scale reads 2.18

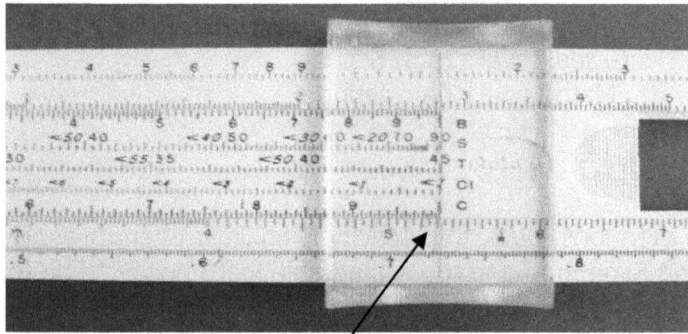

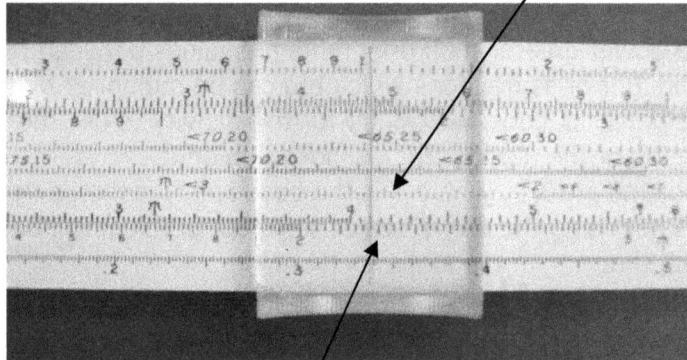

> ➢ Also Note : Keep the 2 in front of the decimal in mind. It actually factors into our answer! It will tell us where to put the decimal!
> ➢ This 2.18 however is not the answer. We need to look this value up on the L scale.
> ➢ You say, wait a minute, that scale is linear and runs from 0 to 1!
> ➢ We look up the decimal to our answer, the .18,
> ➢ Reading from the L scale of .18 to the above D scale which reads 1.51
> ➢

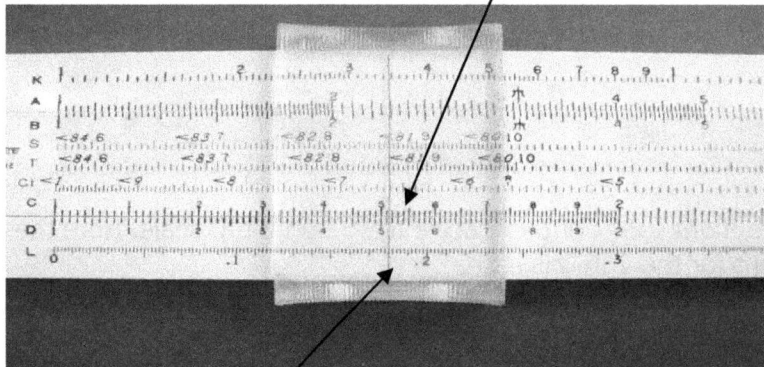

> ➢
> ➢
> ➢
> ➢
> ➢

- We have not forgotten the 2, the 2 tells us the power of 10 for our answer. $10^2$ = 100.
- Hence we move our answer's decimal place twice to the right. Our answer is 151.
- To check this answer one easily needs a scientific calculator and not a regular one. I used one of the fancy ones and found an answer of 151.03011---
- So if the original problem had 3.40, our final answer would match the calculator at 151.
- This is the power of the L scale which allows us to find the powers and roots of any number. Go and try some for yourself! : )

# Activity #9
## The Least Number of Scales Activity
Grade Level : High School
Math Level : Challenging

*An exploration of roots, powers, changing bases and a look at the full power of only using as few as 2 ( or 3 ) Scales on a Slide Rule*

This Activity notes an interesting mathematical exercise for those inclined ( teachers, students, and math geeks alike ) would be to examine different common formulae and test how to solve it with the common 9-Scale and then see if it can be done with only the 2 or 3 Scale version or only using these Scales. Note, that Tables, such as tangent and sine may be needed for some formulae. The best way to approach this Activity is to select a mathematical expression or two and decide how few Scales are needed to accomplish it on a Slide Rule.

A question can arise with a Slide Rule – just how many Scales are needed to perform calculations. As the years went by in the development of the Slide Rule, the number seemed to continually increase. By the mid 1900s and later ( range 1940-1975 ), the number reached around 30!

Surprisingly, most needed calculations in everyday life can be done with 1 Scale ( with 2 Cursors ) or if only one cursor, 2 Scales ( the classic C & D Scales ). This is because most things are linear ( the first power ) and involve basic multiplication and/or division despite magnitude. Also, even if provided the sides of a triangle, though the angle itself cannot be determined, it is often not needed as a result. Many formulae may have the tangent or sine of an angle involved, such as in using vector components and the like. The C & D Scales create the ratios that are the formulae for these trigonometric formulae when needed. Along with that the C & D Scales set up all the similar ratios and are ideal for proportions when doing common calculations and conversions ( see Everyday Life Calculations, Rations & Proportions, and other ratio-based Activities ). This use can be anything such as : miles per gallon, cost per unit ounce, converting grams into ounces, finding decimal equivalents for fractions, and scaling dimensions for similar physical objects or recipe requirements.

The question to ask is, then, what is the next best Scale(s) to add? Just one is needed. The L Scale! Having a C Scale and an L Scale in combination allows not only for all of the basic calculations, but the ability to raise any number to a given power ( 2, 3, 4 et al ), even if that power is not a whole number ( 2.4, 3.1, 5.5 et al ). Also the inverse of this operation, taking roots, of any value, square roots $\sqrt{\phantom{x}}$, cube roots $\sqrt[3]{\phantom{x}}$, and even values like the 5.3$^{rd}$ root is possible with these two scales ( three if there is a C & D Scale with the L Scale ).

**Roots & Powers :**

It is best to start with a simple example to illustrate the point. Let's try $\sqrt{3}$. This is $3^{1/2}$. When examined with a simple Slide Rule with only a few Scales, particularly the L Scale, the log(answer) = log($\sqrt{3}$ ). This means that we have to find the log of the value in question, multiply by its root or power using the C & D Scales. Take that number and look it up on the L Scale and find its antilog on the D Scale for our final answer.

This is from this property of Logarithms :

$$Log ( A^N ) = N*log(A)$$

In this case it can then be seen as 0.5*log(3). Look up the log(3), it is 0.477. Take this value and on the C & D Scales divide 0.477 by 2 which appears between 0.238 and 0.239. With quick mental math it is easy to see it to be 0.2385. Now look up on the L Scale 0.2385 ( as best that can be determined ). Read from the L Scale to the D Scale and find the square root of 3, which turns out to be a bit more than 1.73 ( check it on a calculator ).

Try something more complex, like the 2.3$^{rd}$ root of 7, or $7^{(1/2.3)}$. The same process happens here. The log(7) is 0.845. 0.845/2.3 from the C & D Scales yields 0.367. Taking this from the L Scale back to the D Scale gives a final answer of 2.33 approximately. ( A graphing calculator yields 2.3304*** ).

How about raising to a power? The same process is done for roots as it is with powers. With powers it is easier since typically they are whole numbers and one only has to take the number in question being raised to a power and look up its log value. Multiply this by the power in question and look up the decimal portion of the answer on the L Scale. The digits values in front of the decimal determine the whole number power of 10 to multiply by.

For example what is $4^6$? The log of 4 is 0.602, multiplied by 6 is 3.612. Reading 0.612 on the L Scale and finding the corresponding value on the D scale it is just before 4.1 ( this would be a good rounded off value here ). So 4.1 x $10^3$ = 4,100 is our approximate answer. The actual answer is 4,096. Like the prior example of non-whole number roots, the same can be done with non-whole number powers if desired.

Having this Scale combination ( C, D, & L ) allows for fast conversion of values into their respective log values. This can be useful for evaluating the product and division rules for the law of logarithms ( Log(A*B) = Log(A)+Log(B) ) ( see Logarithmic Product Activity ). Here it can readily be seen that the log of the product of any two values is the sum of the logs of the individual values. Also the construction of a logarithmic set of values for graphing purposes to evaluate the power relation between two variables ( this can be done with any relation where there is a power relation between two variables, such as in The Inquisitive Pioneer ( vol I ) see the Kepler Activity & Rate of Cooling Activity ). Here plotting a log-log graph or a log-linear graph allows one to find the power relation between two variables in question. In these cases values are taken from the D Scale and the log of them are found on the L Scale as needed.

More than this, this simple Scale combination also allows for the evaluation of the Changing of Bases as well as other Laws of Logarithm considerations in Mathematics.

First consider the structure of the logarithm ( used in completing the solution ) :

$\text{Log}_{\text{Base}}$ (Number) = Exponent Value

$\text{Base}^{\text{(Exponent Value)}}$ = Number

Formula for Changing Bases

For any positive numbers, N, A, and B with A ≠ 1 and B ≠ 1,

$$\log_A N = \frac{\log_B N}{\log_B A}$$

Let's say we wanted to determine $\log_2$ (19). The L Scale is set up in base 10, however. Using the formula above, we can evaluate the expression by doing this :

$$\frac{\log_{10}(19)}{\log_{10}(2)}$$

The log(19) from the L Scale is 1.279. Note that there is a 1 in front of the decimal since it is the log(10*1.9) = log(10)+log(1.9). The log(2) from the L Scale is 0.301. With these two values in mind, we use the C & D Scales and divide them. Recognize that we can only estimate to at most 3 significant figures.

$$\frac{1.279}{0.301} \sim 4.25$$

This means that the answer is $\log_2(19) \sim 4.25$, which is $2^{4.25} \sim 19$.

Let's extend the original question to this : $\log_2(19) = \log_3(X)$ where we change bases and want to know what number equals the $\log_2(19)$ if it is the $\log_3$ of some unknown number?

Take our original answer $4.25 = \log_3(X)$

This means $X = 3^{4.25}$

To solve this with the C, D, L system, look up the log(3). This is 0.477. Multiply 4.25*0.477 on the C & D Scales. This turns out to be approximately 2.03. Look up 0.03 on the L Scale and find 1.07 on the D Scale. Since there is a 2 in the ones digits place, this means we multiply this value by 100 and arrive at an approximate answer of 107, which is X. ( A graphing calculator yields 106.6 ) Our final answer : $\log_3(107) \sim \log_2(19)$

Though the majority of Slide Rule history was for more and more Scales, in the background there were a few Slide Rule manufacturers that kept it as simple as possible. There were some, like the Otis King L Model and the Fowler Jubilee ( see slide rule picture page for comparisons and these photos of these models from that page below ) that had only these Scales. Not only did these limit themselves to the fewest of Scales, but they employed very long C Scales. In the case of the Otis King it is 66 inches in length and the Fowler Jubilee Magnum is 76 inches in length. Of course, any other Slide Rule, such as a common 9-Scale version ( K, A, B, S, T, C1, C, D, L ) has them, but in some ways the others are just quicker means to do the work that a simple few could do ( the C, D, L ). Buttons on calculators are meant to make fast work when we know what we are doing, how to do it, and perhaps even why. The Slide Rule is the thinking person's tool and is for those who actively do the math in their mind, the birthplace of ideas, creative thoughts, and insight. It is a fun and challenging adventure to see how few Scales are needed to accomplish a given goal. The user is forced to look at the problem, break it into pieces, have a working knowledge of mathematics – its rules, properties, et al, and employ the solution in a series of organized steps to reach the final destination. This is why examining a Slide Rule is a worthwhile idea to explore in the classroom and for the mathematically curious. Activities such as those noted here in the article and the one in this article illustrate the creative capacity of the Slide Rule and its uses. The key to this idea is this : Find or create various mathematical expressions that can be evaluated and see first which scales will help you solve this problem and then try to consider the problem with fewer and fewer scales to the least number needed to solve a given problem! Enjoy! : )

# Activity #10
## Population Study for Biology
Grade Level : High School
Math Level : Calculating

One of the important things to know about a given species in a given place is their population. Having an idea of how many there are can help biologists and other environmental scientists and conservationists come to understand the overall health of a given species, its needs ( such as food, area, and other factors ) and therefore determine with time and measurement the changes and possible causes to the changes in population size.

First recognize that there is almost no way to count all individuals concretely. Hence statistical sample methods need to be employed to estimate the population. Many safeguards are put in, such as having a number of surveyors taking samples, different times and even methods of sampling and the like so that a reasonable estimate can be reached.

The population of immobile things is straightforward. Since plants are relatively immobile sampling techniques can involve literal counting in a coordinated fashion of marked off areas. Multiplying the number of possible small areas in a given large area by the determined average number of members in a given small area one can arrive at an answer.

This then begs the question, how can one determine the size of a population, say of animals and/or insects since they are mobile? In this case, like the former, a marked off area is swept by the surveyors who collect members of the species in question in a coordinate and programmed method. These are counted and marked and let loose to return to the system they came from. After a prescribed time ( depending on the mobility of the species ) the surveyors return and collect a second sample. The second set has a number that is found and amongst them there may be members of the original capture who are marked. These values are recorded. Using a statistical formula from population sampling, the estimated population size can be determined.

In this activity you are looking for items that will be right in front of you, but this will be done to demonstrate the process and its level of accuracy. You can use hard candies, like M & Ms ( any variety ), or other somewhat regular shaped small candies, such as skittles, chocolate-covered peanuts or raisons, et al. Of course, have permission to use the candy as an experiment and also be aware of allergies in using materials. Here you are the Surveyor trying to count the number of candy 'insects' living in a region. With some modifications, the analysis can be more extensive looking at types with unique marks ( such as represented by the color of the M & Ms ). With this candy activity, be sure to take the same amount of time in doing each of your trials.

Some of the critical assumptions being made here are these :

1) The population is closed ( see #4 )
2) Your sampling process is random
3) All members of the population have the same probability of being captured
4) There are no changes in population from the first to the second sampling – that is no births, deaths, immigration or emigration
5) The marked individuals will spread randomly throughout the population
6) The marked ones are no easier to find than the non-marked ones as well as the surveyor is perfect – that is no marked members are lost or overlooked in the process

**Purpose :** By using the Capture-Mark-Recapture Method the population of a given 'insect' can be determined for a given area.

## Materials :

- Candies that can be marked with an edible marking or non-harmful ( to humans ) marking ( small cut or hole that is identifiable ) – this includes those noted above ( M&Ms, chocolate-covered peanuts, et al ),
- Confectioners 10x baking sugar made into a marking gel with a little water,
- Ruler,
- String,
- Cookie sheet or other flat rectangular pan,
- Slide Rule

## Procedure :

1) Set up : Dump the candies on the rectangular pan surface and spread them out a bit by shaking and just dumping. They do not ( and should not ) have to be uniformly distributed.

2) Use a pan where the ruler can be placed across it. This will act as your survey area. To demark it use string on either side of the ruler so as to note the area that will be surveyed.

3) It is best to have as many measurements as possible, but realize in the real world you may not have the time nor resources to do that many samples. Here 3 is a minimum, 4 is good and 5 is better.

4) You as the Surveyor : At each trial start point place the ruler across the pan. The area it spans is the survey area you would have covered walking straight across.

5) Under the ruler, count the number of candy 'insects' present and mark each one. Use the string to mark the area you are surveying.

6) Note : mark both sides with the edible icing ( easier to find and read ) and be sure to let them dry on wax paper for some time. Touch to test before doing the second time of measurement when the group is put back together and redistributed ( acting like motion of the population of candy 'insects' ).

7) Each of the candy 'insects' is counted and recorded ( S ) and marked with the edible icing.

8) The collected candy 'insects' are put back.

9) Once done, pour the pan of candy 'insects' back into a bag and then redistribute them once again.

10) You as the surveyor go back through the area in exactly the same manner    ( it can be along a different lines ) and collect as many candy 'insects' as can be found moving through in a continuous straight line for the 3-4-5 lines like you did before.

11) The total number collected this time is recorded ( T ).

12) Of the second run collection, the number of marked insects is counted and recorded as well ( M ).

13) All captured candy 'insects' are returned to the 'wild' pan and are still good.

14) Use the data collected to determine the estimated population of candy 'insects' in this area ( P ).

15) Compare the calculated value to the known value to see how close these are as well as calculating the standard error of the for this calculation.

16) For those doing statistical analysis, there is also a standard deviation formula present in the list of calculations that can be performed as well.

## Data :

| Samples Averages | S | T | M |
|---|---|---|---|
| 1 | | XXXXXXXXXXX | XXXXXXXXXXX |
| 2 | XXXXXXXXXXX | | |

S: Is the number of members caught in the first sample and these are
   marked and released.
T : Is the number of members caught in the second sample
M : Is the number of marked members in the second sample

## Calculations :

Be sure to use a Slide Rule for all calculations!

P : Is the estimated Population size

$$\frac{S}{P} = \frac{M}{T}$$

$$P = \frac{S*T}{M}$$

Standard Error :

$$SE = P*(\frac{P-S\ *(P-T)}{S*T*(P-1)})^{1/2}$$

The SE value when added and subtracted from 'P' gives a range that has
a 95% confidence level that the population estimate is in this interval.

The smaller the numbers of members caught (S) and the number of
recaptures ( T ) will generate a larger standard error value.

## Conclusion :

Now examine the actual population as compared to the calculated
value and determine why these are different. Also examine each of the
trials separately to consider why each of these yields such different
results.

Additional Studies :

With M & Ms, you can act as if they are a single population but with
 distinct characteristics ( such as color as in the insect world ) and conduct
 the experiment to determine the proportion of each of these
 characteristic types in the group ( for example orange or green M & Ms ).
 These characteristics can represent wing size or shape, color, size,
 gender, age group, and the like within a population. Follow the same
 basic procedure above, only here keep track not only of marked candy
 'insects' but also their color.

If examining different species or different subgroups in an environment,
 compare the numbers of one subset to another and see how they relate
 to the actual numbers in that group ( the varieties of colors of M & Ms ).

For both experimental designs, if you do the experiment repeatedly along
 different paths and perform separate calculations, then the first
 conclusion point is to examine the various calculations as compared to
 each other to see how similar or different these estimates are and why
 one might speculate as to the cause of this.

Alternative Outdoor Method :
( no eating these, but make it safe for the outdoor animals that encounter
your 'bugs' )

This same activity can be done outdoors using nut 'insects' ( actually
 environmentally safe nuts that birds can eat ) to estimate the
 population of them out there. In essence this is a representative
 activity, but it is representative of the real thing. Here, the surveyor
 needs a parent or teacher who knows the population and where they are
 at and acts as the distributor of the materials. Also these materials are
 not eaten. They are merely captured, marked with an environmentally
 safe marker or mark, and then redistributed randomly ( by the parent or
 teacher ) and the surveyor goes through again ( using time limits and
 counting along a given path ). The method you are employing is in the
 capture process, you are trying to find and capture as many members
 from this population as you can in your path through the known
 population area in a specific time. To randomize their motion, a teacher
 or parent acts as the distributor and the person who knows the actual
 population size to begin with so that you as the surveyor are trying to
 find an unknown. To make the activity more motivating to look for
 environmental causes that affect populations, utilize 2 different areas
( say near and under a tree as compared to open grass areas ) where
 each has a uniquely different population. The surveyor may speculate as
 to why this is so.

<u>Other Studies</u> :

Though not this experiment, it is easy to run statistical computations of
  M&Ms, such as the probability of taking a green M & M out of the bag
  and then another one ( or some other sort of question. Simply know the
  total amount in question, the numbers of each of the group members
  ( number of green, red, orange, et al ) and determine the probability of
  each separately. Ask various questions and employ basic rules of
  probability for determining the outcome. ( For more information on the
  formulae, look at the Statistics with a Slide Rule Activity on the website
  www.cosmicquestthinker.com ).

# Activity #11
## Experiments with Sprouting Seeds :
Grade Level : Middle School
Math Level : Calculating

Simply put, seed plants are vascular plants that produce seeds. Seeds in plants are essentially the whole plant at its earliest stage. The three basic components are the seed coat, the food supply and the embryo. The embryo, like other living things, has the DNA that has its code for its whole self. In essence, in the seed is the whole of a tree, even one that is hundreds of feet tall one day. Seeds are a great adaptation of plants so as to be able to disperse their offspring. Seeds can be of many different sizes depending on the given plant and their size can be part of how they are transported and how far they travel when produced by the parent plant.

Depending on the plant, its seed can have a given number of adaptations to deal with environmental factors – such as temperature changes, too much or too little water, and the like. Some seeds can even be consumed by some animals and excreted but are not digested along the way and hence are carried far from the parent plant and can grow where they land. Seeds can come from plants that can be for trees, flowers, bushes, and crops. There are many plants grown from seeds, both by us and just in nature itself. Think of seeds planted for corn, tomatoes, carrots, potatoes, along with flowers of all sorts as well as trees. Humans even use seeds for food like many insects and small animals as well. Think for example of pumpkin seeds, sunflower seeds, and the like. Since there are so many predators that feast on seeds this is all the more reason that plants produce large numbers of them so as to ensure survival.

Despite being a rather self-contained and sustaining system, seeds do need a reasonably nutrient-rich soil system, adequate water supply, and when the plant sprouts and grows adequate light to continue to full growth. Each of these environmental factors can affect the seed and its ability to grow. In our Activities that follow, we explore many different ideas. Read through the following directions first since these are the basis of all the Activities.

## Important General Guidelines for All of the Activities for Seeds / Plants :

1. First have parental permission and help in each of these Activities undertaken.
2. It is important to realize that in these Activities they are genuine science investigations and NOT plants to be considered for consumption. Do not eat or sample these experimental plants at any time.
3. In using seeds make choices that best fit your needs. The suggestions listed here are not the only possibilities. Do your homework and do some research online. For parents in considering seeds – the larger the seeds the better for handling, also look for ones that sprout quickly and have fast growth rates.
4. Though general ideas are given here – be sure to follow the directions with the seeds you purchase such as for watering and the like.
5. To be consistent in amounts use measuring spoons in watering. Be careful in experiments where there are different watering bottles for each of the groups – be sure to clean the measuring spoon between each of the watering of each of the groups.

6. The noted tools for time are the Clock and the Calendar for each of the Activities and this is obvious. The clock is merely to keep track of the amount of time passing when needed, such as the number of hours of light on the plant – if it is called for. The Calendar is used to schedule things such as watering and measuring.

7. Best to have the Activities indoors ( though the last one involves outdoor ones since it involves sunlight, but realize that this can be done indoors too! With the plants near a window facing the Sun – this is best since all other variables other than light would be controlled ). With indoor Activities it is easiest to control the various variables, such as lighting / temperature changes / et al.

8. It is best to go soil-less for experiments looking at factors in the water when watering the plants (such as salt or acid), by using paper towels (or coffee filters or cotton balls) as the 'soil' when it comes to seeds.

9. If using plants and soil is a must, then it is best to by a potting soil instead of using soil from outside due to various things that can be in it ( bacteria, et al ).

10. Realize that many factors can affect your seeds, their sprouting and their growth. Try carefully to control and monitor each of the factors that can have an impact on their success.

11. In the case of each of the Trials : a Trial is a set of seeds/plants that have identical conditions under consideration. The number in a set should be no less than 2 but can and should be 3 to 5 for best results. Whatever the number in a set in a Trial, that is the number of measurements you will have to make for a given Trial and then this value will be Averaged so as to represent the results for the given Trial. Note : If a seed/plant fails or dies along the way, clearly its measurements will not change anymore, but it is still recorded and factored in.

12. There are two types of Trial Groups : The Control Group which is s a set of plants ( can be only 1, but best to have a number of 2 – 5 ) – this group receives all of the same basic controlled factors that the Treatment Group receives, but none of the treatment ( the variable being examined in the experiment ). The Treatment Group, which are all plants that receive treatment. In the case of the Treatment Group, there are subsets, such as plants #1-3 receive the maximum treatment, plants #4-6 receive half of the treatment, and so on. The number of subsets depends on what the experiment is about, the amount of resources, time, and the like.

13. When handling seeds/plants, be sure to have clean hands – in fact, if possible use plastic gloves. This is so not to transfer bacteria or cause damage. This is the reason for allowing the seeds to spout and grow in a visual and soil-less environment of a clear plastic cup and the seed is in a folded paper ( paper towel or coffee filter ) or cotton balls where the seed can visually be seen against the cup as it grows and sprouts – there is no handling once planted.

14. A flexible measuring tape is best to use to measure since it can follow the path of the plant. Realize you are going to have a certain level of estimation in your measurements – also abiding with the idea of not handling the sprouting seeds or plants when and where possible. In the matter of the spouting seeds in plastic cups, do not remove them to measure at any time. So be sure to position them to be visible.

15. There is an important realization in measurements. There are two things springing from the seed pod – the stem and the root. We are measuring both in total. So you measure from a given point on the seed and through the entire length of the sprout as it will be referred to, which really encompasses both the stem and the root.

16. Recommendations for Seeds to Use : This list does not list them in any particular order and always consider the amount of time, the conditions you have available, the experiment you have chosen, the size of the seeds, and even the cost of them. Also – even if they are normally seeds used for items that are food, here they are merely experiments and NOT to be eaten at any time.

17. Type of Seeds :
   a. Various Gourd Seeds,
   b. Birdhouse Gourd Seeds,
   c. Luffa Sponge Gourd Seeds,
   d. Beet Seeds,
   e. Turnip Seeds,
   f. Radish Seeds,
   g. Cosmos Seeds,
   h. Sunflower Seeds,
   i. Blue Lake Bean Seeds,
   j. Pencil Wax Bean Seeds,
   k. Cucumber Seeds,
   l. Poppie Seeds,
   m. Marigold Seeds
   n. Et al

18. Recommendation : Though plastic cups are used, be sure to reuse them for other plants in the future, or rinse and reuse for other projects, or if done with them, then recycle – thanks.

## Soilless Seed Environment Photo

## Activity 1 for Seeds
## Comparison of the Sprout Time and Growth Rate of the Chosen Seeds in Soil vs. a soilless environment ( paper towel ) Activity

Though this Activity reports itself to be an investigation of growing plants without soil, we are using some sort of medium ( paper towel, coffee filter, ? ), but do note the fact that growing plants without soil is not a new idea nor is it impossible.

Clearly one of the first things one notes needed by plants is soil ( along with light and water ). In the case of the soil it has a basis for support, contains the nutrients and is where the root systems obtain water as well.

With the increase in understanding the needs of plants even as early as the late 1800s it was reasoned that one could use a nutrient-rich solution imitating the soil environment enjoyed by seeds and plants. The key to the success in growing plants without soil is being able to know what is needed by the seeds. Some soilless studies involve sand, others use gravel, and some involve water. This water-based process is called hydroponics.

Many of these ideas may be fruitful in times ahead so as to increase the world's food supply. In our Activity we are mainly studying not so much the effect of a lack of soil, but asking the question : what benefits do the soil have? Do the seeds in the soil sprout more frequently, earlier, and/or grow faster than in a condition without the soil?

**Purpose :** Comparison of the Sprout Time and Growth Rate of the Chosen Seeds in Soil
vs. a soilless environment ( paper towel or other )

## Materials :
- Flexible Measuring Tape,
- Paper, Pencil ( to keep records ),
- Choose one of the following mediums to act as the soilless material : 1 ) Paper Towel sheets ( several sheets ), 2) Coffee Filters ( several ), 3) Cotton Balls (amount depends on the number of cups used ),
- Water,
- Measuring Spoon ( size will be determined by the amount of water needed and the watering schedule determined by the directions on the seeds chosen ), ( NOTE : it is often a small one, such as 1/3 or ¼ tsp and it is done every other day or so for example )
- Clear or semi-clear plastic cups ( number needed depends on Group sizes – each seed has its own cup in both the soil and soilless groups ),
- Regular bag of potting soil,
- Chosen Seed type from seed list above,
- Paper Towel tubes or Toilet Paper Towel tube(s),
- Scissors,
- Clock and Calendar,
- Slide Rule

## Set Up, Basic Operations and Procedure :

1. First Read the General Guidelines in the opening of this list of Activities and follow these.
2. First choose the seeds to be used. Read their directions carefully for watering and care.
3. Be sure to have chosen your soilless medium ( see Materials List above ).
4. Each cup, soil or soilless ( control or treatment groups ), will have the same general arrangement : Cup with medium on outer rim, seed positioned so that it can be seen, and the central region of the cup occupied by a portion of paper towel or toilet paper tube so that the amount of medium ( soil or paper or cotton ) is kept to a minimum ( see photo ).
5. NOTE : The use of the tube is only to reduce the amount of soil and/or soilless medium materials. However, if you have parental permission and adequate materials to fill the cup with the media ( soil and soilless ) then the tubes are not needed. Realize that one cannot have one of the media with a tube and the other not with it however, since there is a lack of consistency from one group to the other.
6. If tubes are chosen, parents should cut these to fit the cups.
7. The clock is used for the amount of time per day that the plants receive light ( see Data section on directions for a Table ) and to record the time of day one waters ( as well as the amount ). The calendar is used to note the number of passing days – Day 0 is when you begin and this can be any given day and date on the calendar.
8. The Day noted does not have to be sequential ( ie 0, 1, 2, etc ), but realize one should check on the seeds daily and give a quick visual look since you are looking for when each sprouts individually and this needs to be recorded! You can choose a non-sequential time frame, for example every 2 or 3 days is sufficient ( 0, 3, 6, … ), but note measurements are taken then for all.
9. 
10. How to Measure :
11. 
12. There are actually several measurements being taken :
13. The first is to note how many plants 'succeed', that is sprout, and how many 'fail' or do not sprout. Keep track of this tally. You will use it to compute the probability of success ( see Calculations ).
14. If you are keeping track of various variables, such as watering amounts, and the like, these can be graphed as well ( bar graph recommended ). Also the amount of light in a given day can be measured as well. – If interested in the numbers you can calculate average amounts ( water, light, et al per day too ).
15. 
16. Measuring the Plants :
17. 
18. Once a seed has sprouted ( note it is situated in the cup and in the medium ( towel or other ) so that it can be seen ), you have already chosen a time frame for measurement – it can be daily, but you can measure every 2 or 3 days – no matter the choice, be consistent.

19. Use the flexible measuring tape start with the end at the same point on the seed and unfurl it to the length of the sprout following it like a line. Pinch the measuring tape where the sprout ends, then read it to the nearest $1/10^{th}$ of a centimeter. Record these results.
20. How to deal with sprouting seeds where the sprout grows away from the side of the cup and is hard to measure? – Use a plastic stirring stick or straw if available to prod them to be visible – careful not to damage them.
21. Continue your experiment until you have chosen to end the observations ( typically no longer than 2-3 weeks past the time of initial sprouting ).
22.
23. Procedure for the Experiment
24.
25. In this experiment we are measuring the rate of growth of the sprouting seeds of choice in an environment with soil and one without soil.
26. As noted above, keep track of the measurements of the seed lengths throughout the experiment for both the soil and soilless environs. Keep separate tables. Be sure each plant in each subgroup has its own column – perhaps mark with a letter or number so as not to mistake them.
27.
28. Once all measurements are taken :
29. First determine the frequency of success for each subgroup : Take the number of sprouted plants in a given subgroup divided by the total number of seeds in the subgroup.
30. If wanted you can determine the average amount of water per day or light per day : For example, sum up all the amount of hours and minutes of sunlight that the plants experienced and divide by the total number of days ( to determine hours per day ) of sunlight on the plants.
31. The main calculation : Determine the average rate of growth by formula : Sum up the final measures of the plants in a subcategory ( soil, soilless ) and divide by the number of plants in the subgroup.
32. You can also determine the average rate of growth by graphing it. On a x-y axis graph days on the x-axis and the measure of length for a given sprout on the y-axis. For a given plant connect the dots by drawing the 'best fit' straight line and then determine the slope of this line. With each slope for a given subgroup, sum these up and divide by the number of plants in that given subgroup to find the average.
33. Now consider the question as to whether the plants grew 'better' or not in a soilless environment or one with soil?

## Data :

Conditions for All Plants ( both the Control Group and the Treatment Groups ) :

1. As in most of these Experiments, all the plants should be indoors and in the same general location – this ensures the same temperature, lighting, and air circulation conditions.

2. Each is placed in a window so that each set receives the same amount of light daily. Note – it is a good idea to record the amount of time the plants receive light ( start and stop times to the day ) so that this variable can be looked at in case of any influence in results after the experiment. – A table for this is not provided, but easily made : It is only 2 columns ( assuming each of the plants in the Control and Treatment Groups is treated the same ), the first column is the Day 0, 1, et al, and the second column is the recorded amount of sunlight or light. This is why we have a clock being used. What is the start and stop time, and the amount of time therefore for light incident on these plants.
3. Though the seeds in both the paper towel and soil have been placed along the clear or semi-clear cup boundary so as to be visible, it is best to keep them turned away from the direct sunlight ( if sunlight is used ) or whatever light source is employed. The key is that all receive the same light from the same source.
4. Each receives the same amount of water on the same watering schedule. This will depend on the type of seed used. Be sure to look at the directions on the package for the seeds. Best to use a measuring spoon in dispensing the water to ensure consistency in amount. – Note – best to have a note spelling out the schedule so as to keep on it. You can also create a table for this as you did in step 2 for the table for amount of light, only here the second column is the watering schedule and when you did it on what day.
5. Measurements are taken in the same time intervals for all plants – but note the importance of checking daily so as to see if any of the seeds have sprouted so that the first day can be noted and which day it was. Best to use a letter and a number to indicate the start, such as X3 means the seed sprouted and did so on the 3$^{rd}$ day, for example – you make up your own code and define it.

**Control Group :** Plants in soil environment

| Day | Plant 1 : Length of Sprout | Plant 2 : Length of Sprout | Plant N : Length of Sprout |
|---|---|---|---|
| 0 | 0 | 0 | 0 |
| | | | |
| | | | |

- Note : 'Day' does not need to be sequential and can be multiples if needed ( 2 or 3 days for example )

**Treatment Group :** Plants in soilless environment

| Treatment Group | Plant Range Numbers | Treatment |
|---|---|---|
| A | Plants # 1-3 | Paper Towel as its soil |

- Note : The number of Plants in a given Treatment Group is up to you, the values listed above are just a suggestion.

Treatment Group A Data :

| Day | Plant 1 : Length of Sprout | Plant 2 : Length of Sprout | Plant 3 : Length of Sprout |
|---|---|---|---|
| 0 | 0 | 0 | 0 |
| | | | |
| | | | |

- Note : Each Treatment Group has to have its own Data Table.

## Calculations :

Be sure to use your Slide Rule!

**Slope :**

$$m = \frac{\Delta Y}{\Delta X}$$

**Average :** Is the sum of all of the measurements divided by the number of measurements

$$X_{ave} = \frac{\sum_{i=1}^{n} x_i}{n}$$

**Frequency of Success :** The decimal value ( out of 1.0 ) for the rate of successful sprouts and/or plants growing ( s ) in a given Trial ( n )

$$f = \frac{s}{n}$$

## Conclusion :

How similar or dissimilar are the sprouting rates and growth rates for the seeds in the soil versus the ones in the paper towel environment? Did one of the two have a higher success rate? – What sort of factors do you think may account for these differences and can you think of a way to construct an experiment to measure this? What of the rates of growth of the Stem and the Root – were they similar or not?

## Activity 2 for Seeds

Comparison of the Sprout Time and Growth Rate of the Chosen Seeds in a neutral environment vs. progressively acidic environments

Though we think of rain as only water, and it mostly is just that, rain can have other materials in the water as well. This is because of the nature of the water molecule, which is often called the universal solvent – that is to say, many things can dissolve in it.

Acid Rain forms naturally even without human intervention, but the activities of humans to put into the atmosphere compounds which dissolve in the water vapor in the air cause it to acidify so that it along with other naturally occurring compounds increase the level of acidity of the rain when it falls to Earth. Primarily it is sulfur and nitrogen compounds forming these acids. What does it mean to be acidic? Essentially materials can be acidic, basic or rather neutral in terms of their behavior. The level of acidity is measured by the pH scale. Water is considered neutral and has a value of 7.0 on a logarithmic scale of 0 to 14. On this scale the acid values range from 0-6 while basic are from 8-14 for the most part. Interestingly, rain tends to be slightly acidic with a typical pH range of 5.3 to 6.0. Acid Rain has pH values less than 5.3. To be acidic is a measure of the reactivity of a substance and its ability to be a $H+$ donor – the lower the pH value the more $H+$ ions available. Each whole number value in the pH scale is a 10 fold increase or decrease ( depending on the direction on the scale ) from the adjacent number.

Acid Rain can fall on all surfaces and has a number of effects. It can also enter streams and rivers and effect the biomes in the water systems. Acid Rain has its greatest effects on aquatic systems since it can affect the pH of the water system. When the pH of a water system drops too low it can affect both plant and animal life to the point of making the body of water in question incapable of supporting any life whatsoever.

Acid Rain on plants on the land can also have a number of impacts. Acid Rain falling in a forest can damage tree bark, make a tree lose its leaves – both of which impact a tree to make it vulnerable to harsh weather, disease, and insects. The younger the plants, the greater the overall impact on them. The Acid Rain depletes the soil of nutrients, since the acidic water can interact with various materials in the soil. The lower pH values can also kill some of the soils microorganisms. The damaged soil can also lead to accelerated erosion.

Interestingly Acid Rain is not only wet rainfall, but also includes dry deposition ( dust mainly ) that carry these nitric and sulfuric compounds. Also rainfall is not only rain, but can be fog, mist, and snow as well.
Acid Rain was noted in the late 1800s with the increase in industrialization but no major movements in society or government occurred until the 1960s.

These are the reasons to explore the effects of acids on plants in this Activity. We are not exploring the acids in question here, but will use things like lemon juice or vinegar to substitute in our exploration of how acidic solutions affect the sprouting rate and growth rate of a given plant that you have chosen to examine.

While measuring this, think of ways we can reduce exhaust, such as from cars and the like which strongly contributes to these nitrogen and sulfur containing compounds so as to reduce acid rain and its effects. Also what types of materials can act as what are called buffering agents to neutralize these effects ( hint : you might want to run an experiment where you use baking soda in your 'soil' when watering it with an acidic solution ).

**Purpose :** Comparison of the Sprout Time and Growth Rate of the Chosen Seeds in a neutral environment vs. progressively acidic environments

## Materials :

- Chosen Seed type from seed list above,
- Flexible Measuring Tape,
- Paper, Pencil ( to keep records ),
- Paper Towel sheets ( 2-3 ),
- Clear or semi-clear plastic cups,
- Distilled or Tap Water,
- Lemon Juice and/or Vinegar ( can try two different acids if you have the time and option – be sure to treat each separately ),
- pH measuring strips,
- Measuring Cups,
- Measuring Spoon ( size will be determined by the amount of water needed and the watering schedule determined by the directions on the seeds chosen ), ( NOTE : it is often a small one, such as 1/3 or ¼ tsp and it is done every other day or so for example )
- Pop Bottles,
- Paper Towel or Toilet Paper Towel tube(s),
- Scissors,
- Clock and Calendar,
- Goggles,
- Slide Rule

## Procedure :

1. First Read the General Guidelines in the opening of this list of Activities and follow these.
2. First choose the seeds to be used. Read their directions carefully for watering and care. Watering here is critical since it contains the key variable under consideration, the amount of acid. Be consistent in timing and amount, but of course each group has its own watering bottle. – Best to use a measuring spoon in watering, but clean it between groups, of course.
3. Be sure to have chosen your soilless medium ( see Materials List above ).

4. Each cup, is soilless ( both the control ( non-acidic ) or treatment groups ( acidic ) ), will have the same general arrangement : Cup with medium on outer rim, seed positioned so that it can be seen, and the central region of the cup occupied by a portion of paper towel or toilet paper tube so that the amount of medium ( soil or paper or cotton ) is kept to a minimum ( see photo ).

5. NOTE : The use of the tube is only to reduce the amount of soilless medium materials. However, if you have parental permission and adequate materials to fill the cup with the media ( paper towel or whatever is being used ) then the tubes are not needed. Realize that one cannot have one of the media with a tube and the other not with it however, since there is a lack of consistency from one group to the other.

6. If tubes are chosen, parents should cut these to fit the cups.

7. The clock is used for the amount of time per day that the plants receive light ( see Data section on directions for a Table ) and to record the time of day one waters ( as well as the amount ). The calendar is used to note the number of passing days – Day 0 is when you begin and this can be any given day and date on the calendar.

8. The Day noted does not have to be sequential ( ie 0, 1, 2, etc ), but realize one should check on the seeds daily and give a quick visual look since you are looking for when each sprouts individually and this needs to be recorded! You can choose a non-sequential time frame, for example every 2 or 3 days is sufficient ( 0, 3, 6, … ), but note measurements are taken then for all.

9.

10. How to Measure :

11.

12. There are actually several measurements being taken :

13. The first is to note how many plants 'succeed', that is sprout, and how many 'fail' or do not sprout. Keep track of this tally. You will use it to compute the probability of success ( see Calculations ).

14. The key measurement you are keeping track of is the watering – be sure to know the directions on the seed packages and do some research. It is best to keep the medium damp. Though acid is being used, it is common household materials, vinegar or lemon juice and is watered down so in this one case you can touch the soilless media to see if damp and needs to be watered. However, it is best to not do this since in a regular lab this is never done. It is best to go by visual inspection and the watering directions. Any further assistance should be by your parents.

15. Be sure to have separate watering bottles ( see Procedure below ) and use small amounts – the same amount in the same time intervals for each of the plants, for example a tablespoon every 3 days ( read the directions for the seeds ).

16. There may be other variables, such as the amount of light being kept track of, and the like, these can be graphed as well ( bar graph recommended ). – If interested in the numbers you can calculate average amounts ( water, light, et al per day too ).

17.

18. Measuring the Plants :

19.

20. Once a seed has sprouted ( note it is situated in the cup and in the medium ( towel or other ) so that it can be seen ), you have already chosen a time frame for measurement – it can be daily, but you can measure every 2 or 3 days – no matter the choice, be consistent.
21. Use the flexible measuring tape start with the end at the same point on the seed and unfurl it to the length of the sprout following it like a line. Pinch the measuring tape where the sprout ends, then read it to the nearest $1/10^{th}$ of a centimeter. Record these results.
22. How to deal with sprouting seeds where the sprout grows away from the side of the cup and is hard to measure? – Use a plastic stirring stick or straw if available to prod them to be visible – careful not to damage them.
23. Continue your experiment until you have chosen to end the observations ( typically no longer than 2-3 weeks past the time of initial sprouting ).
24.
25. Procedure for the Experiment
26.
27. In this experiment we are measuring the rate of growth of the sprouting seeds of choice in an acidic environment as compared with a non-acidic environment.
28.
29. How to create the Acidic Solution to treat the plants with :
30.
31. The first thing to note is that we are not preparing a strong acid, but lab protocol is an absolute necessity. You must wear your goggles in the preparation of the acidic solution. Next, when it is created, acids are poured into water and not vice versa.
32. First determine how many acidic treatment groups you are going to have : it is recommended to have 2 – one that has a slight acidity ( about half as much ) and one that is much more acidic ( about twice as much ).
33.
34. How to create the acidic watering solution to use :
35.
36. It is best to have an idea of how long the experiment will run ( 2-3 weeks ) so that you can determine the amount of water you need to set aside – though with the directions you can refill it once empty.
37. It is best to use cleaned out regular size pop bottles – one for the water, one for the slightly acidic water, and the one with the most acidic water.
38. Fill an 8 oz measuring cup ( 1 cup ) with water and with a funnel put this amount into each of the pop bottles.
39. Next use a ½ cup and put this amount of water into the one marked water only.
40. Use the ½ cup and fill it with your parentally approved and supervised acid – vinegar is recommended and place this into the highly acidic water bottle.
41. Now use the ¼ cup and fill it with water and then pour it into the ½ cup.
42. Now use the ¼ cup and fill it with the chosen acid and pour it into the now half-full ½ cup and then pour this into the slightly acidic water bottle.
43. Use the pH paper to measure the pH of the solution and record this value.
44.
45. Measuring the Plants through time :

46. As noted above, keep track of the measurements of the seed lengths throughout the experiment for both the acidic and non-acidic environs. Keep separate tables. Be sure each plant in each subgroup has its own column – perhaps mark with a letter or number so as not to mistake them.
47.
48. Once all measurements are taken :
49.
50. First determine the frequency of success for each subgroup : Take the number of sprouted plants in a given subgroup divided by the total number of seeds in the subgroup.
51. If wanted you can determine the average amount of water per day or light per day : For example, sum up all the amount of hours and minutes of sunlight that the plants experienced and divide by the total number of days ( to determine hours per day ) of sunlight on the plants.
52. The main calculation : Determine the average rate of growth by formula : Sum up the final measures of the plants in a subcategory ( acidic, non-acidic ) and divide by the number of plants in the subgroup.
53. You can also determine the average rate of growth by graphing it. On a x-y axis graph days on the x-axis and the measure of length for a given sprout on the y-axis. For a given plant connect the dots by drawing the 'best fit' straight line and then determine the slope of this line. With each slope for a given subgroup, sum these up and divide by the number of plants in that given subgroup to find the average.
54. In conclusion, consider the original question stemming from the hypothesis – how does an increasing acidic environment affect the rate of plant growth?

## Data :

Conditions for All Plants ( both the Control Group and the Treatment Groups ) :
1. Same Climate ( light, temperature, humidity, atmosphere ),
2. Same watering schedule ( though the same time and amounts, the only difference is the solution used however )

**Control Group :** Plants only receiving pure water

| Day | Plant 1 : Length of Sprout | Plant 2 : Length of Sprout | Plant N : Length of Sprout |
|---|---|---|---|
| 0 | | | |
| | | | |
| | | | |

**Treatment Group :** All Treatment Groups receive a prescribed amount of acidic solution in their watering.

| Treatment Group | Plant Range Numbers | Treatment |
|---|---|---|
| A | Plants # 1-3 | |
| B | Pants # 4-6 | |
| C | Etc | |

- Note : The number of Plants in a given Treatment Group is up to you, the values listed above are just a suggestion.

**Treatment Group A Data :**

| Day | Plant 1 : Length of Sprout | Plant 2 : Length of Sprout | Plant 3 : Length of Sprout |
|---|---|---|---|
| 0 | | | |
| | | | |
| | | | |

**Treatment Group B Data :**

| Day | Plant 4 : Length of Sprout | Plant 5 : Length of Sprout | Plant 6 : Length of Sprout |
|---|---|---|---|
| 0 | | | |
| | | | |
| | | | |

- Note : Each Treatment Group has to have its own Data Table.

## Calculations :

Be sure to use your Slide Rule!

**Slope :**

$$m = \frac{\Delta Y}{\Delta X}$$

**Average :** Is the sum of all of the measurements divided by the number of Measurements

$$X_{ave} = \frac{\sum_{i=1}^{n} x_i}{n}$$

**Frequency of Success :** The decimal value ( out of 1.0 ) for the rate of successful sprouts and/or plants growing ( s ) in a given Trial ( n )

$$f = \frac{s}{n}$$

## Conclusion :

Is there an optimal environment for these types of seeds in terms of an acidic environment? Do you think all seeds react this way – if not, why not try another experiment with permission and supplies, of course?

## Activity 3 for Seeds

Comparison of the Sprout Time and Growth Rate of the Chosen Seeds in salt-free soil environments vs. increasingly salty soil environments

Though one might not think so at first, but soil is a mixture of many things, and salts of all sorts are part of the soil. Interestingly some salt is essential for plants as well as animals. The key to the environment of the plants, however, is the fact that plants are more affected by their environment due to their relative immobility as compared to animals and insects.

Too much of anything, even if it a necessity, can be a bad thing ( much like the other extreme of too little of a necessity ). In the case of salt, if too much enters the soil environment of the plants it increases what is called the salinity of the soil. Most of the increase in salinity in soil is due to natural weathering of rocks. Soils like this are found in semi-arid and arid regions. Since there is little rainfall here, the salts in the upper regions of the soil cannot be moved through the soil. These salts then can leach water from plants and root systems of plants.

Too much salt in the environment of a plant can cause damage, water loss, and inhibited processes. Salt is commonly sodium chloride ( NaCl ) and too much sodium, as in humans, can lead to damage and even death in the case of plants.

This increase in salinity is not only natural, however, and has some impact by humans by doing things such as affecting the water table in a given area, and the like. Lands that are becoming too saline-based are on the increase and the need for farming lands is increasing greatly due to the increases in human populations.

In this Activity we are exploring a range of salt-solutions watering our plants and noting the rate of sprouting changes and any changes in the rate of growth of the plants that are chosen. Note that some plants do fair better than others when it comes to salt, but it is known that most crops are salt-sensitive. This may be an area to explore and consider in the future when looking to new areas to grow foods when looking at possible cross-breeding or genetic manipulation of plants in the future.

**Purpose :** Comparison of the Sprout Time and Growth Rate of the Chosen Seeds in salt-free soil environments vs. increasingly salty soil environments

## Materials :

- Chosen Seed type from seed list above,
- Flexible Measuring Tape,
- Paper, Pencil ( to keep records ),
- Paper Towel sheets ( 2-3 ),
- Tap Water,
- Table salt ( NaCl ),
- Measuring Cups,
- Measuring Spoons ( size will be determined by the amount of water needed and the watering schedule determined by the directions on the seeds chosen ), ( NOTE : it is often a small one, such as 1/3 or ¼ tsp and it is done every other day or so for example )
- Clear or semi-clear plastic cups,
- Paper Towel or Toilet Paper Towel tube(s),
- Pop Bottles,
- Scissors,
- Clock and Calendar,
- Slide Rule

## Procedure :

1. First Read the General Guidelines in the opening of this list of Activities and follow these.
2. First choose the seeds to be used. Read their directions carefully for watering and care. All should be watered in the same manner and amount – the only difference here in this experiment is the watering bottle is different for each group due to having salt or not having salt. Best to use a measuring spoon to water each of the groups with from the respective watering bottles – but be sure to clean the spoon between groups.
3. Be sure to have chosen your soilless medium ( see Materials List above ).
4. Each cup is soilless ( the control group is salt-free while the treatment group(s) have increasing amounts of salt – good to have at least 2 perhaps 3 of progressive amounts of salt groups ), will have the same general arrangement : Cup with medium on outer rim, seed positioned so that it can be seen, and the central region of the cup occupied by a portion of paper towel or toilet paper tube so that the amount of medium ( paper towel, coffee filter or cotton ) is kept to a minimum ( see photo ).
5. NOTE : The use of the tube is only to reduce the amount of soilless medium materials. However, if you have parental permission and adequate materials to fill the cup with the media ( soilless ) then the tubes are not needed. Realize that one cannot have one of the media with a tube and the other not with it however, since there is a lack of consistency from one group to the other.
6. If tubes are chosen, parents should cut these to fit the cups.

7. The clock is used for the amount of time per day that the plants receive light ( see Data section on directions for a Table ) and to record the time of day one waters ( as well as the amount ). The calendar is used to note the number of passing days – Day 0 is when you begin and this can be any given day and date on the calendar.

8. The Day noted does not have to be sequential ( ie 0, 1, 2, etc ), but realize one should check on the seeds daily and give a quick visual look since you are looking for when each sprouts individually and this needs to be recorded! You can choose a non-sequential time frame, for example every 2 or 3 days is sufficient ( 0, 3, 6, ... ), but note measurements are taken then for all.

9.

10. How to Measure :

11.

12. There are actually several measurements being taken :

13. The first is to note how many plants 'succeed', that is sprout, and how many 'fail' or do not sprout. Keep track of this tally. You will use it to compute the probability of success ( see Calculations ).

14. If you are keeping track of various variables, such as watering amounts, and the like, these can be graphed as well ( bar graph recommended ). Also the amount of light in a given day can be measured as well. – If interested in the numbers you can calculate average amounts ( water, light, et al per day too ).

15. The watering is recommended to be measured since this has the variable, ie saltiness, that is being examined.

16.

17. Measuring the Plants :

18.

19. Once a seed has sprouted ( note it is situated in the cup and in the medium ( towel or other ) so that it can be seen ), you have already chosen a time frame for measurement – it can be daily, but you can measure every 2 or 3 days – no matter the choice, be consistent.

20. Use the flexible measuring tape start with the end at the same point on the seed and unfurl it to the length of the sprout following it like a line. Pinch the measuring tape where the sprout ends, then read it to the nearest $1/10^{th}$ of a centimeter. Record these results.

21. How to deal with sprouting seeds where the sprout grows away from the side of the cup and is hard to measure? – Use a plastic stirring stick or straw if available to prod them to be visible – careful not to damage them.

22. Continue your experiment until you have chosen to end the observations ( typically no longer than 2-3 weeks past the time of initial sprouting ).

23.

24. Procedure for the Experiment

25.

26. In this experiment we are measuring the rate of growth of the sprouting seeds of choice in an environment with soil and one without soil.

27. As noted above, keep track of the measurements of the seed lengths throughout the experiment for both the soil and soilless environs. Keep separate tables. Be sure each plant in each subgroup has its own column – perhaps mark with a letter or number so as not to mistake them.

28.
29. How to create the salt-solutions to water the plants with :
30.
31. The first question to answer is how many Treatment Groups do you plan to examine? Minimally one is required. If time and materials permit 2 or 3 are quite good. For the sake of creating solutions 2 will be considered where one is a little salty and the other is quite a bit more salty.
32. Use as many cleaned out regular pop bottles as needed for solutions. There is one for pure water, one for the small salt concentration and one for the large salt concentration, and so on. Be sure to label them and be careful in watering so as to be consistent and watering the proper group with the proper solution.
33. Use a 8 oz measuring cup and a funnel. Fill the measuring cup with water and funnel into each of the watering bottles 8 oz of water.
34. Next into the small salt concentration place one teaspoon of salt. Put on the cap and shake to mix.
35. Next into the large salt concentration place one tablespoon of salt. Put on the cap and shake to mix.
36. If not all of all of the salt does not mix, then add the same amount of water to each of the solution bottles until all of the salt dissolves in the large salt concentration bottle.
37. If the experiment continues and the bottle runs out, then refill in the same manner as before.
38.
39. Once all measurements are taken :
40. First determine the frequency of success for each subgroup : Take the number of sprouted plants in a given subgroup divided by the total number of seeds in the subgroup.
41. If wanted you can determine the average amount of water per day or light per day : For example, sum up all the amount of hours and minutes of sunlight that the plants experienced and divide by the total number of days ( to determine hours per day ) of sunlight on the plants.
42. The main calculation : Determine the average rate of growth by formula : Sum up the final measures of the plants in a subcategory ( water, low salt concentration, large salt concentration ) and divide by the number of plants in the subgroup.
43. You can also determine the average rate of growth by graphing it. On a x-y axis graph days on the x-axis and the measure of length for a given sprout on the y-axis. For a given plant connect the dots by drawing the 'best fit' straight line and then determine the slope of this line. With each slope for a given subgroup, sum these up and divide by the number of plants in that given subgroup to find the average.

## Data :

Conditions for All Plants ( both the Control Group and the Treatment Groups ) :
1. Same Climate ( light, temperature, humidity, atmosphere ),
2. Same watering schedule ( though the same time and amounts, the only difference is the solution used however )

**Control Group :** Plants only being watered with pure water

| Day | Plant 1 : Length of Sprout | Plant 2 : Length of Sprout | Plant N : Length of Sprout |
|---|---|---|---|
| 0 | | | |
| | | | |
| | | | |

**Treatment Group :** Plants being watered with salty water.

| Treatment Group | Plant Range Numbers | Treatment |
|---|---|---|
| A | Plants # 1-3 | |
| B | Pants # 4-6 | |
| C | Etc | |

- Note : The number of Plants in a given Treatment Group is up to you, the values listed above are just a suggestion.

**Treatment Group A Data :**

| Day | Plant 1 : Length of Sprout | Plant 2 : Length of Sprout | Plant 3 : Length of Sprout |
|---|---|---|---|
| 0 | | | |
| | | | |
| | | | |

**Treatment Group B Data :**

| Day | Plant 4 : Length of Sprout | Plant 5 : Length of Sprout | Plant 6 : Length of Sprout |
|---|---|---|---|
| 0 | | | |
| | | | |
| | | | |

- Note : Each Treatment Group has to have its own Data Table.

## Calculations :

Be sure to use Your Slide Rule !

**Slope :**

$$m = \frac{\Delta Y}{\Delta X}$$

**Average :** Is the sum of all of the measurements divided by the number of measurements

$$X_{ave} = \frac{\sum_{i=1}^{n} x_i}{n}$$

**Frequency of Success :** The decimal value ( out of 1.0 ) for the rate of successful sprouts and/or plants growing ( s ) in a given Trial ( n )

$$f = \frac{s}{n}$$

## Conclusion :

How did the saltiness of the watering solution affect the rate of plant growth? Do you think all seeds react the same way? Though soil was not used here, do you think an environment with soil would be different – why or why not?

## Activity 4 for Seeds

Comparison of the Growth Rate of sprouted the Chosen Seeds in sunlight vs. artificial light environments ( as well as environments of short time in light vs. long time in light )
Prelude :

The 3 basic things that plants seem to need are : soil, water, and light. We have examined in the prior Activities a situation without soil and the effects of affected water ( too acidic and having too much salt ). In this Activity we explore one or both possible roads : the type of light and/or the amount of light.
For a given plant the amount of light needed varies, so there is no universal answer as to how much is needed. As with most things in life, too much or too little of a necessity can cause harm, and in the case of light this is true. Too little light can cause plant growth to be rather sparse, the plants weak with a wider separation of the leaves. Too much light can cause the plants to dry out and even cause leaf edges to have a burnt out appearance. Too much light can cause a weakened and drooping plant too.
Whatever plant seeds you have chosen it is best to do some research on the recommended amount of light needed. This value can be a guide to your Activity as to how much you might want to explore in terms of lighting for your plant. Obviously we cannot control the amount of sunlight, so it is best to use an artificial lighting system. It does not have to be a special lamp for plants, just use a conventional lamp.


**Purpose :** Comparison of the Growth Rate of sprouted the Chosen Seeds in sunlight vs. artificial light environments ( as well as environments of short time in light vs. long time in light – if you have two of the same lamps )

## Materials :

- The Chosen seeds from the list above and are already sprouted into plants ( for each sample set it is best to have no less than 2, but 3 to 5 is best ),
- Flexible Measuring Tape,
- Tap Water,
- Measuring Spoon ( size will be determined by the amount of water needed and the watering schedule determined by the directions on the seeds chosen ), ( NOTE : it is often a small one, such as 1/3 or ¼ tsp and it is done every other day or so for example )
- Small cups ( one for each plant – even those outside if used ),
- Incandescent Desk Lamp and/or Florescent Desk Lamp (Note – Can do both separately ) ( note : if you have two of the same type of lamps then can do long term versus short terms exposure to light ),
- Paper Towel or Toilet Paper Towel tube(s),
- Scissors,
- Clock and Calendar,
- Slide Rule

## Procedure :

1. First Read the General Guidelines in the opening of this list of Activities and follow these.
2. First choose the seeds to be used. Read their directions carefully for watering and care.
3. Be sure to have chosen your soilless medium ( see Materials List above ).
4. Each cup is soilless ( for both the control ( sunlight ) and treatment groups ( artificial light ) ), will have the same general arrangement : Cup with medium on outer rim, seed positioned so that it can be seen, and the central region of the cup occupied by a portion of paper towel or toilet paper tube so that the amount of medium ( soil or paper or cotton ) is kept to a minimum ( see photo ).
5. NOTE : The use of the tube is only to reduce the amount of soil and/or soilless medium materials. However, if you have parental permission and adequate materials to fill the cup with the media then the tubes are not needed. Realize that one cannot have one of the media with a tube and the other not with it however, since there is a lack of consistency from one group to the other.
6. If tubes are chosen, parents should cut these to fit the cups.
7. The clock is used for the amount of time per day that the plants receive light ( see Data section on directions for a Table ) and to record the time of day one waters ( as well as the amount ). The calendar is used to note the number of passing days – Day 0 is when you begin and this can be any given day and date on the calendar.
8. The Day noted does not have to be sequential ( ie 0, 1, 2, etc ), but realize one should check on the seeds daily and give a quick visual look since you are looking for when each sprouts individually and this needs to be recorded! You can choose a non-sequential time frame, for example every 2 or 3 days is sufficient ( 0, 3, 6, … ), but note measurements are taken then for all.
9. How to Measure :
10. There are actually several measurements being taken :
11. The first is to note how many plants 'succeed', that is sprout, and how many 'fail' or do not sprout. Keep track of this tally. You will use it to compute the probability of success ( see Calculations ).
12. If you are keeping track of various variables, such as watering amounts, and the like, these can be graphed as well ( bar graph recommended ).
13. In this experiment the amount of light in a given day is to be measured as well ( both for artificial groups and the control group with sunlight ).
14. Another calculation to consider once the experiment is done from the numbers is this : you can calculate average amounts ( water, light, et al per day too ).
15.
16. Measuring the Plants :
17.
18. Once a seed has sprouted, its date is noted. If you like you can monitor the rate of growth of the seedling while out of the light. As to the measurement routine at this point ( note it is situated in the cup and in the medium ( towel or other ) so that it can be seen ), you have already chosen a time frame for measurement – it can be daily, but you can measure every 2 or 3 days – no matter the choice, be consistent.

19. Use the flexible measuring tape start with the end at the same point on the seed and unfurl it to the length of the sprout following it like a line. Pinch the measuring tape where the sprout ends, then read it to the nearest $1/10^{th}$ of a centimeter. Record these results.

20. How to deal with sprouting seeds where the sprout grows away from the side of the cup and is hard to measure? – Use a plastic stirring stick or straw if available to prod them to be visible – careful not to damage them.

21. Note – You do not have to measure the sprouting seeds however. Instead you can wait until the seeds have penetrated the soil ( or soilless ) environment and then measure with a flexible measuring tape on a scheduled time frame ( 2 or 3 days, for example – be sure to be consistent ) and be careful not to damage the growing plants.

22. Continue your experiment until you have chosen to end the observations ( typically no longer than 2-3 weeks past the time of initial sprouting ).

23.

24. Procedure for the Experiment

25.

26. In this experiment we are measuring the rate of growth of the sprouting seeds of choice in an environment with sunlight and an environment with the same amount of light only it is artificial light.

27. It is best even on overcast days to still have the artificial light on the treatment group since there is light, though not at full intensity, on the control group.

28. At this point you can also have two experiments running at the same time, with adequate materials available : Another 1 or more treatment groups that receive artificial lighting only for shorter or longer times of exposure than the initial treatment group that matches the amount of sunlight exposure.

29.

30. How to create the conditions for the artificial and sunlight groups :

31.

32. Clearly each should be in the same room so that all of the other environmental variables, including the watering schedule are the same.

33. The sunlit one is in the window while the artificial lamp one ( or more ) is nearby at a safe distance from the lamp being used.

34. The treatment group plants with artificial light should have materials to act as barriers when needed to prevent light from reaching the plants – this can include construction paper, newspaper, a box that can act as a lld for the entire group in the treatment group, et al. Be sure not to disturb them as this can affect their growth.

35. It is best to have at least a treatment group that matches the sunlit group in terms of time in the light. Other considerations can be one with less light (try 50%, for example) and one with more light ( 1.5x the amount of sunlight ). A very good control to have is one set of seeds totally in the dark receiving no light at all ( hint ).

36. Be sure to have each group carefully labeled and timed correctly.

37. Be sure not to look into the lamp, touch the lamp, and employ other common sense behaviors when using the lamp.

38. Note that the lamp can be for all treatment groups since there are barriers to cover them as needed.

39.

40.  As noted above, keep track of the measurements of the seed lengths throughout the experiment for both the soil and soilless environs. Keep separate tables. Be sure each plant in each subgroup has its own column – perhaps mark with a letter or number so as not to mistake them.

41.

42.  Once all measurements are taken :

43.

44.  First determine the frequency of success for each subgroup : Take the number of sprouted plants in a given subgroup divided by the total number of seeds in the subgroup.

45.  If wanted you can determine the average amount of water per day or light per day : For example, sum up all the amount of hours and minutes of sunlight that the plants experienced and divide by the total number of days ( to determine hours per day ) of sunlight on the plants. – Here the light per day is recommended due to the nature of this experiment.

46.  The main calculation : Determine the average rate of growth by formula : Sum up the final measures of the plants in a subcategory ( sunlit, artificial light ) and divide by the number of plants in the subgroup.

47.  You can also determine the average rate of growth by graphing it. On a x-y axis graph days on the x-axis and the measure of length for a given sprout on the y-axis. For a given plant connect the dots by drawing the 'best fit' straight line and then determine the slope of this line. With each slope for a given subgroup, sum these up and divide by the number of plants in that given subgroup to find the average.

48.  Examine results for patterns and conclusions as to whether sunlight or artificial light had a greater effect on the rate of plant growth. Also, if other experiments such as the amount of light were done, examine the rate of growth as compared to shorter and/or extended periods of artificial light illumination.

## Data :

Conditions for All Plants ( both the Control Group and the Treatment Groups ) :
   1.  Same Watering, Soil, and Environmental Conditions ( other than light )

## Control Group : Plants exposed to sunlight

| Day | Plant 1 : Length of Sprout | Plant 2 : Length of Sprout | Plant N : Length of Sprout |
|---|---|---|---|
| 0 | | | |
| | | | |
| | | | |

**Treatment Group :** Plants exposed to artificial light

| Treatment Group | Plant Range Numbers | Treatment |
|---|---|---|
| A | Plants # 1-3 | |
| B | Pants # 4-6 | |
| C | Etc | |

- Note : The number of Plants in a given Treatment Group is up to you, the values listed above are just a suggestion.
- Note : Be sure to record the wattage and type of light used as well. Perhaps this can be a further experiment where different wattages and/or types of lights of the same wattage are used for comparison.

Treatment Group A Data :

| Day | Plant 1 : Length of Sprout | Plant 2 : Length of Sprout | Plant 3 : Length of Sprout |
|---|---|---|---|
| 0 | | | |
| | | | |
| | | | |

Treatment Group B Data :

| Day | Plant 4 : Length of Sprout | Plant 5 : Length of Sprout | Plant 6 : Length of Sprout |
|---|---|---|---|
| 0 | | | |
| | | | |
| | | | |

- Note : Each Treatment Group has to have its own Data Table.

## Calculations :

Be sure to use your Slide Rule !

**Slope :**

$$m = \frac{\Delta Y}{\Delta X}$$

**Average :** Is the sum of all of the measurements divided by the number of measurements

$$X_{ave} = \frac{\sum_{i=1}^{n} x_i}{n}$$

**Frequency of Success :** The decimal value ( out of 1.0 ) for the rate of successful sprouts and/or plants growing ( s ) in a given Trial ( n )

$$f = \frac{s}{n}$$

## Conclusion :

How do the rates of growth compare between natural and artificial light? Did you also try an experiment of increasing time of light exposure and how do these growth rates compare? Does there seem to be a maximum value where plants do the best versus less or more light? Do you think all seeds react the same way? Are there ways of increasing light exposure without power use?

# Activity #12
## Rate of Respiration in Yeast Activity
Grade Level : High School
Math Level : Calculating

Yeast come from one of the main Kingdoms in Biology ( Monera, Protista, Fungi, Plantae, Animalia ), namely the Fungi. It is a living organism.

There is a variety that is used in various cooking enterprises of people, such as making breads and pastries where they release carbon dioxide which, in turn, causes the breads and pastries to rise.

This release of carbon dioxide is a part of a process called Respiration. Respiration is done not only be yeast, but all other life forms ( such as bacteria, animals, and even plants ). It is done in order to obtain energy. It involves the breakdown of nutrients and the release of carbon dioxide and the energy.

Respiration is a critical part of the Carbon Cycle as well.

The chemical equation for respiration in our yeast experiments is :

Glucose + Oxygen yields Carbon Dioxide + Water + Energy ( Heat )

$C_6H_{12}O_6$ + $6O_2$ yields $6CO_2$ + $6H_2O$ + Energy

Our Activity explores the amount of nutrient supplied to the yeast ( namely the sugar ) and its effect on respiration indirectly.

In the case of the Activity, not only do we measure the amount of $CO_2$ given off ( as captured in a balloon ) to act as a measure of respiration.

**Purpose :** To measure by inference the rate of respiration of yeast using controlled amounts of sugar in a water medium at a constant temperature through changes in volumetric changes of the emission of carbon dioxide

**Materials :**

- Active Dry Yeast,
- Water,
- Regular Table Sugar ( $C_6H_{12}O_6$ ) or use packets of sugar ( note – it is pre-measured ),
- Number of Regular Plastic Pop Bottles,
- Sewing Flexible Tape Measure ( if none, use string for balloon circumference and measure on a meter stick ),
- Measuring Set of Teaspoons and Tablespoon,
- Balloons ( that can go over bottle lip, round thin latex variety is best ),
- Masking Tape,
- Timer or Stopwatch,
- Slide Rule

Note : Do not use Quick-rising Yeast, instead used the type in baking breads )

**Procedure :**

1. **Activity : Amount of Respiration with changing sugar amounts**
2. Follow this procedure as many times as you have trials for the Activity involving the changing amount of sugar.
3. The key to these trials, which can be simultaneously ran ( since sugar is the variable of interest, but you need that number of pop bottles for different trials )
4. For all the trials, the variables that remain constant are :
5. The temperature of the water solution ( use tap water, and each bottle with be in the same environment – at room temp ( just sitting out ) ), in the refrigerator with ample space, in a cooler that has its ice changed with some regularity, or even a cool basement  ) ( T )
6. The amount of water used ( Measured by measuring cups and poured into the pop bottle via a funnel ) ( M )  : 8 oz ( 1 cup ).
7. The amount of Yeast used in each trial ( Y ) : 1 tsp.
8. A good control is to have all the same, but without the sugar at all for both the thermometer measuring situation as well as one with a balloon on it ( there should be no changes to either of these ).
9. The amount of time : This depends on when measurements are made. They can be made every hour, two hours, 6 hours, 8 hours, 12 hours, even up to 24 hours apart for up to 3 days if one wants. Time is the critical factor affecting this experiment.
10. In each of the trials in Activity, the variable that changes is the amount of sugar ( S ) used. It should be done in incremental amounts ( 1 tsp, 2 tsps, 3 tsps, ( or use sugar packets as a good choice ), et al for as many trials that are done ).

11. Note : It is best to seal the balloon to the bottle with masking tape.
12. Note : In the case of any and all bottles for the experiments ( here and later ), it may be a good idea to swirl it a bit lightly so as to mix the components and allow them to interact.
13. Every Time Unit chosen ( ½ hr, 1 hr, 2 hr, 6 hr, 8 hr, ½ day, or full day ) the bottle with the balloon has its circumference measured and recorded ( C ) in the data table.
14. Note that after each trial, the pop bottle must be thoroughly rinsed for the next trial ( Again Note that each of these trials can be done at the same time, only one needs that amount of supplies, pop bottles being one of the critical ones here ).
15. Calculations :
16. For the Balloon Experiment : Convert each of the Circumferences into Diameters ( d ).
17. For each set of results ( they are based on the amount of sugar used ) Graph the Independent Variable ( Time ) on the x-axis and the Dependent Variable ( Diameter d ) on the y-axis where each line is for a given sugar amount.
18. For the graph, draw a best fit line for each sugar amount and calculate slope.
19. If you suspect that the relationship is not linear, try a log-log plot of the variables and find the best fit line slope of this to determine exponential relations.
20. Note : Once done with materials, exercise care in removing the balloon.


**Data :**

    **Activity :** Amount of Sugar Used : _____ ( unit used )

    Note : For varying amounts of Sugar Used, each requires its own table as presented below. All have the same initial conditions. Be sure to decide ahead of time the number of trials to be ran ( how many different amounts of sugar used – best to do it as increasing ).

    Measurement of Changing Balloon Circumference over Time
    ( Yeast Respiration indirectly )

| Time ( unit chosen – hours, etc ) | Balloon Circumference ( cm ) |
|---|---|
| 1st | |
| 2nd | |
| 3rd | |
| 4th | |
| 5th | |
| ... | |
| Last | |

## Calculations :

Be sure to use your Slide Rule!

Slope of Graph :

$$m = \frac{\Delta Y}{\Delta X}$$

Diameter ( d ) of a Circle from Circumference ( C ) :

$$C = \pi * d$$

## Conclusion :

What were the rates of change of the dependent variable with time for your Activity? Does it remain constant? Why or why not?

## Interesting Alternatives :

In all cases, the balloon, bottle, yeast, water, and sugar are used.

Try a comparison Activity of types of sugar ( artificial, sugar in natural juices such as apple juice as compared to cane sugar ). Be sure to have the same conditions for all of the trials, such as the amount of water used, the amount of yeast, et al.

Another consideration is the pH level of the environment. Start with several pop bottles with 8 oz. of water, 1 tsp of yeast, and 2 tsps of sugar. The control bottle has nothing added ( other than additional water up to 10 tablespoons ), but to two of the bottles, add 5 tablespoons of vinegar ( and 5 tablespoons of water ) to one of them, and the next add 10 tablespoons of vinegar. The vinegar will create an 'acidic' environment. Compare and contrast these rates of respiration.

A final consideration might be the temperature of the water initially. Have a control group at room temp while another pop bottle with the balloon sits in an ice bath in a bucket and perhaps a third in a refrigerator of sufficient size. Compare these rates.

# Activity #13
## Conservation of Mass Activity
Grade Level : High School
Math Level : Challenging

**Conservation Laws** are the cornerstone of the sciences. In Physics, Chemistry, and Biology ( plus all of the other interrelated sciences ), a conservation law is one where a certain measureable property of an isolated system does not change as the system evolves. The characteristic in question is defined within a 'closed system' ( there is no transfer across the boundary to this system ), the amount of that measured characteristics ( mass, energy, momentum ) does not change within it. That characteristic can transfer this amount to other elements of the system ( and in the case of energy other types of energy ) but the total amount at the outset is the same as at the closure of the situation.

There are even two major categories of conservation laws. There are ones that are referred to as exact laws ( they have never been shown to be violated ) and the others as approximate laws ( which are true in particular situations – such as low speeds, short time scales, or certain interactions ).

In the **exact conservation laws** category there are these :

- Conservation of Energy
- Conservation of Linear Momentum
- Conservation of Angular Momentum
- Conservation of Electric Charge
- Conservation of Color Charge
- Conservation of Weak Isopin
- Conservation of Probability

In the **approximate conservation laws** category are these :

- Conservation of Mass ( non-relativistic speeds )
- Conservation of Baryon Number
- Conservation of Lepton Number ( Standard Model )
- Conservation of Flavor
- Conservation of Parity
- CP Symmetry

The Conservation Laws are actually relatively recent additions to the realm of Science. Basically stated in their proper application – the total quantity in question ( say for example momentum ) for a given set of objects in a closed system ( one in which no external forces enter – say friction, et al ) then the quantity remains constant. The one under consideration here is the Conservation of Mass.

Though it falls in the 'not exact' category, it is because of two basic reasons. First, this idea applies to non-relativistic situations. This essentially means speeds much less than the speed of light. This is actually the norm for most matter in the universe, including large objects like planets and stars, so there is not too much of a problem there. Second, it holds for when it involves 'ordinary' chemical reactions. This basically means those events not involving radioactivity.

Where the law comes from, historically is from early chemists who in performing measured experiments of materials before and after a reaction came to find that the amount of the materials involved in these chemical reactions remains constant, even when the products were not the same molecular form as what one started with. Take, for example a reaction of hydrogen gas and oxygen gas. When starting with twice as much hydrogen gas in volume as oxygen, and having a complete reaction, all of it becomes the more common $H_2O$ or liquid water. In terms of mass, if one starts with 4 g of hydrogen gas and 32 g of oxygen gas, the yield would be 36 g of water (4 g + 32 g).

Though that seems somewhat obvious, it took until the 1700s to have the proper tools and understanding to reach that conclusion. Also realize the difficulty in knowing which substance is which, measuring the mass of a gas, and other problems that may occur as well. What if, for example one were to react two materials and it gave off a odorless and colorless gas. If one did not take care to trap this product the mass of the reactants would be greater than the remaining mass of products in the beaker since the gas had escaped!

The ancient Ionian Greek, **Democrius**, postulated that all things are made of atoms – though he had no fully direct evidence of such. He notes that they are solid, essentially homogeneous, and indivisible. **John Dalton** ( 1766 – 1844 ) creates what is known as **Dalton's Atomic Theory**. This states that matter is composed of very small particles called atoms which are indivisible and indestructible. The atoms of a given element are identical in size, mass, and overall chemical properties. Also the atoms of a given element are unique, hence different than those of other elements. Elements will combine in simple whole number ratios to form compounds. The final part of his idea is the basis of Conservation of Mass and it states that in a chemical reaction, the atoms are merely separated and then recombined or rearranged in the outcome products.

Many of the ideas in his theory stand the test of time to a certain extent, but with some modifications. For example, it is found that for a given element, there are slightly different masses for some of the members of a species of element. This is because some of the members are isotopes ( elements with a different mass number due to the different number of neutrons, but having the same number of protons as the other species members ). It was found that there are 3 fundamental particles that comprise all atoms, independent of species of atom. They are essentially made of a nucleus comprised of protons and neutrons, which in turn, is orbited by a cloud of electrons. The number of protons equals the number of electrons in any ordinary ( hence electrically neutral ) atom. The number of protons turns out to be the distinguishing characteristic for determining the type ( or species ) of atom it is. For example, all Hydrogen atoms have 1 proton, but there are isotopes that have 1 or 2 neutrons as well ( a hydrogen with 1 neutron is called deuterium while a hydrogen with 2 neutrons it called tritium ). Another example is Carbon, where all Carbon atoms have 6 protons and the most common form has 6 neutrons while isotopes ( like carbon 14 ) have a different number of neutrons.

The key to all of these investigations ( done by numerous chemists and scientists ) of the 1700s and 1800s is that they verified Dalton's initial findings of the combinations in whole-number ratios as well as the consistency of the mass of the reactants equaling the mass of the products independent of the materials used.

Further in time, the elements were found to have comparable properties in given families and could be arranged in a table we refer to as the **Periodic Table** today. This came from people such as **Antione Lavoisier** ( 1743 - 1794 ) and later and more extensive work by **Dmitri Mendeleev** ( 1834 - 1907 ).

The importance of these ideas and the conservation of mass cannot be over-emphasized. The whole notion of balancing equations and having a balanced equation is built upon this concept – the idea that the mass of the reactants equals the mass of the products.

The following Activities explored here do not prove the Conservation of Mass, but merely explore it.

In the case of the Activity 1 involving Vinegar ( $CH_3COOH$ ) and Baking Soda ( $NaHCO_3$ ) the reaction is thus :

vinegar + baking soda  yields  sodium acetate + water + carbon dioxide

$$CH_3COOH + NaHCO_3 \text{ yields } NaC_2H_3O_2 + H_2O + CO_2$$

In the case of Activity 2 involving Water and an Alka-Seltzer Tablet, the reaction is thus :

water + sodium carbonate yields water + sodium hydroxide + carbon dioxide

$$H_2O + Na(HCO_3) \text{ yields } H_2O + NaOH + CO_2$$

**Purpose :** To demonstrate the Conservation of Mass from the chemical reaction of vinegar and baking soda with before and after mass measurements of the reactants and the products.

**Purpose :** To determine the percent of mass given off as carbon dioxide gas from the reaction of vinegar and baking soda.

**Purpose :** To determine the percent of the mass of gas in a whole Alka-Seltzer Tablet as calculated from the mass of gas given off by a piece of the tablet.

## Materials :

- **Activity 1 :**
- Measuring Cup ( ¼, 1/3, or ½ cup sizes useful ),
- Vinegar ( $CH_3COOH$ ) ( this is called acetic acid solution 5% ),
- Baking Soda ( $NaHCO_3$ ) ( this is called sodium hydrogen carbonate ),
- Plastic Pop Bottle ( regular size ),
- Balloon ( large enough to fit on pop bottle ),
- Piece of Tissue ( Kleenex ),
- Kitchen Mass Scale ( can measure to 0.1 g and precision of 0.01g best ),
- Measuring Set of Teaspoons ( ¼, 1/3, ½ are best )
- Safety Goggles,
- Slide Rule
- **Activity 2 :**
- Water,
- Alka-Seltzer Tablet ( ),
- Kitchen Mass Scale ( best to use a Triple Beam Balance ),
- Measuring Cup ( approx. 1/3 or ½ cup best ),
- Small Cup ( regular coffee cup or plastic cup best ),
- Safety Goggles,
- Slide Rule

**Notes on Safety :** In order to do the Activity, one must employ safety by following directions, not overdoing it with amounts of vinegar and/or baking soda, wearing safety goggles and above all – have permission and supervision of parents for this.

**Notes :** The greater the precision of the scale, the better the results. Be sure to know both the mass limits of the scale in use as well as the level of precision of the scale. Test amounts ahead of the actual activity when taking measurements.

**Procedure :**

1) For all Activities, be sure to zero out the scale, and do the activity at least twice if not three times in order to examine data for consistent results.
2) Be sure to wear safety goggles for both Activities.
3) **Activity 1**
4) Measure 1/3 tsp of baking soda onto a piece of tissue.
5) Measure 1/3 cup vinegar and pour into the pop bottle.
6) Place the system ( balloon, piece of tissue ( Kleenex ), 1/3 tsp of baking soda, 1/3 cup vinegar ( now inside the pop bottle ), pop bottle ) on the scale and measure its total mass ( g ).
7) For another test, one could take the individual masses and then add them instead of recording the total in the data table for comparison.
8) Loosely wrap the tissue and put into the pop bottle. Quickly cover the pop bottle with the balloon.
9) Shake lightly the bottle so as to continue the reaction of the baking soda and the vinegar. At this point the balloon should be filling up some.
10) Place the reacting system on the scale at this point and measure its mass. Ideally it is the same as the initial mass taken before combination.
11) Carefully remove the balloon and release the carbon dioxide gas into the air ( do not breath this in ).
12) Replace the balloon with the reacted contents inside the pop bottle on the scale once more and take the final mass reading. ( Ideally it will be somewhat less than the initial mass reading – somewhere between 0.1 to 1.2 g depending on the amount of reactants used ).
13) Perform the Activity 2-3 times more. It is best to use the same amounts for at least 2 trials and to perform at least two different sets of initial masses used ( use ½ tsp and ½ cup, for example ) for comparative results.
14) Compute the Percent Gas in the Mixture.
15) Are there any conclusions that you can reach knowing the concept of Conservation of Mass and the data you have recorded?
16) **Activity 2**
17) Measure the mass of a single whole Alka-Seltzer Tablet ( g ).
18) Break a piece off the tablet ( about ¼ to 1/3 is best ) and measure its mass ( g ).
19) Measure 1/3 cup of water into a cup ( small plastic best – note needs to be tall enough so when reaction between water and tablet takes place it does not spill out )
20) Take the mass of the system ( cup with water, piece of tablet ) on the scale ( g ).
21) Place the tablet piece into the water and let it react.
22) Lightly fan it and swirl it a bit until the reaction is totally complete.
23) Again take the mass of the system (now the cup, water, and reacted tablet piece) ( g ).
24) Do this activity 2 or 3 times for comparative results.
25) How similar ( dissimilar ) are the initial and final mass measurements ( they should be different where the initial mass should be greater than the final mass, since in this activity you have let the gas go instead of trapping it as you did in Activity 1 ).
26) Calculate the Percent Gas in the Tablet Piece and the Percent Gas in the Whole Tablet using the relations in the Calculations table.
27) Use the Conservation of Mass to describe your findings and calculations.

## Data :

## Activity 1 :

Mass of Balloon :   _____ g
Mass of Pop Bottle :   _____ g
Mass of Tissue :   _____ g
Mass of Baking Soda : _____ g
Mass of Vinegar :   _____ g

System : Pop bottle, Baking Soda, Tissue, Vinegar, Balloon

Initial Mass of the System :
Total Mass of System before reaction : _____ g

Total Mass of System ( with balloon attached ) after reaction : _____ g

Final Mass of the System :
Total Mass of System ( with detached & deflated balloon ) after the
reaction : _____ g

Change in Mass : _____ g

## Activity 2 :

Mass of Cup :   _____ g
Mass of Water :   _____ g
Mass of Alka-Seltzer Tablet ( whole ) : _____ g
Mass of Alka-Seltzer Tablet Piece :   _____ g

System : Cup, Water, Alka-Seltzer Tablet Piece,

Initial Mass of the System :
Total Mass of System before reaction : _____ g

Final Mass of the System :
Total Mass of System ( with detached & deflated balloon ) after the
reaction : _____ g

Change in Mass : _____ g

## Calculations :

Be sure to use your Slide Rule!

## Statements of Conservation of Mass :

**Initial Mass = Final Mass**

**Mass of the Reactants = Mass of the Products**

## Activity 1 :

**Mass of Gas in Sample = Initial Mass of System − Final Mass of System**

## Percent of Gas in Mass System :

$$\% \text{ Gas} = \frac{\text{Determined Mass of Gas ( g )}}{\text{Total Mass of System Initially ( g )}} * 100\%$$

## Activity 2 :

**Mass of Gas in Sample = Initial Mass of System − Final Mass of System**

*System : Mass of ( Cup + Water + Tablet )*
*Initial : Before the chemical reaction*
*Final : After the chemical reaction*

## Proportion to Determine the Mass of Gas in the Tablet :

$$\frac{\text{Mass of Gas in Whole Tablet (g)}}{\text{Mass of Whole Tablet (g)}} = \frac{\text{Mass of Gas in Sample (g)}}{\text{Mass of Sample Piece of Tablet (g)}}$$

## Conclusion

From Activity 1, how do your Initial Mass readings compare to your intermediary Mass Readings where the balloon is attached to the pop bottle and has a gas in it?

Do you think the percentages of gas would increase, decrease or stay the same as you incrementally increased the amount of the reactants in both Activity 1 and Activity 2?

In the case of Activity 2, if the container had been closed initially, would the same have occurred with regards to the measurements of the total mass of the whole system both before and after the reaction?

# Activity #14
## Pressure & Volume Considerations of Boyle's Law Activity
Grade Level : High School
Math Level : Challenging


Unlike the other common states of matter, liquids and solids, gases have a certain amount of compressibility. The question then arises : Is there a relationship between the volume of the gas and the pressure ( force per unit area ) exerted upon it?

The answer was found over 300 years ago in 1662 and was articulated by the physicist Robert Boyle ( whose assistant was none other than Robert Hooke ) and called Boyle's Law ( sometimes referred to as Boyle-Mariotte Law as it was independently found by Edme Mariotte in 1676 ).

The Law states :

"For a fixed amount of an ideal gas kept at a fixed temperature, Pressure (P) and Volume (V) are inversely proportional"

$$P*V = k$$

$$P_1*V_1 = P_2*V_2$$

In essence, what it says is that as one of these variables increases,( say for example Pressure ), the other will decrease ( here then Volume ). Hence if we graph the relation of Volume to Pressure, we should find an inverse relation. ( $y = \frac{1}{x}$ ).

This also implies that their product ( Pressure x Volume ) must be a constant. Of interest is this product. What are the units of Pressure and Volume combined? It turns out to be Newton*meter. This is known today as a Joule! In essence, the amount of energy of this system remains constant as long as no energy is allowed to leave or enter the system. These ideas would take nearly 200 years to be better understood with the conservation law of energy.

Before that discovery, scientists of the day were trying to find the explanation of the behavior. Even Newton's Laws were invoked to understand this phenomenon. Newton's Principia was published in 1687 and here he showed mathematically in the case of an elastic fluid at rest and if there are interactive forces between the particles that act in accordance to an inverse proportionality to distance, then the density of said fluid would be proportional to its pressure. Though useful, it is not the full mathematical story of the answer to this enigma. This idea requires the later creation of the Kinetic Theory of Gases and the formation of the Ideal Gas Law.

In nature, most gases will act like ideal gases under regular pressures and temperatures. In eras past the 1600s where greater and smaller values in pressures and temperatures could be explored then variation to these relations could be uncovered. This deviation comes to be called the compressibility factor.

Though the Law was crafted not from theory but from experimental grounds, it initially met resistance. Objections were not for the data itself but since the newly formed idea of the Kinetic Theory of Gases also allowed for this outcome. The theory notes that atoms exist, their motions follow Newton's laws, and undergo perfectly elastic collisions, plus some other criteria. Daniel Bernoulli in 1738 even derived Boyle's Law using Newton's Laws of Motion but this was overlooked until 1845 when John Waterston published a paper laying the foundations of the modern kinetic theory of matter. During this century later these ideas are accepted through the work of James Prescott Joule, Rudolf Clausius, and Ludwig Boltzmann ( by 1898 ).

In the end, Boyle's Law, Charles' Law ( another Activity #20 ), and Gay-Lussac's Law when combined become the modern day common Gas Law we all know and use in chemistry. When taken in connection to Avogadro's Law the form can be generalized to the more modern-day and familiar, Ideal Gas Law.
( PV = nRT )

Our goal is to use a contained amount of air ( in a syringe ) and apply increasing pressure to it ( through added weights ) and measure the changes in the volume of the trapped air. From our data, we wish to uncover what Boyle had found for himself over 300 years ago.

**Purpose :** To change the pressure experienced by a measured initial volume of air in order to measure changes in the volume of the air to uncover the relation between pressure and volume at a constant temperature. ( In order to verify Boyle's Law ).

**Materials :**

- large Plastic Syringe ( 60 mL is good ),
- Caliper ( for measuring ),
- Wooden Block ( 3-4 inches length & width, thickness at least 1" ),
- known Weights ( exercise circular are best ),
- Drill & Drill Bit ( see set up ),
- use large Books, et al ( see set up ),
- Goggles,
- Slide Rule

**Safety Procedure :**

If doing this at home, be sure to have parent permission and supervision as power tools are needed here. Always wear your goggles using tools and/or equipment. Without proper supervision or permission or not exercising safety then in any of those cases Do Not do any of the following. This is an Activity that requires parental help and supervision – particularly in doing some of the set up as well as the experiment as well. At all times exercise caution, wear goggles, be sure to have the best available materials and equipment in doing the activity.

## Set Up Preparation Before Activity Procedure :

1) Note : This set up should be done by an adult.
2) In the matter of the weights, it is best that they are circular, disk-shaped and have markings for kilograms. With no metric mass markings, you have to do the conversions.
3) Best choice of weights are masses of 1 kg apiece ( about 2 lbs. ) and have at least 4 of these ( 6 is good ).
4) Use the caliper to measure the syringe 'needle' diameter and find a slightly smaller drill bit than this size. Drill a hole that is as deep as the 'needle' exit of the syringe into the flat of the block, being sure that the wood block is thick enough so as not to come through. Make it thin enough for a tight fit. Do not put it in yet.
5) Measure the diameter of the syringe plunge rubber stopper head. Determine radius from this and place this in your data table.
6) Place the block with the hole face up in a manner on level floor so that it is between large books or boards so that it does not move easily.
7) Note that when the syringe is in this place standing vertically and weights on it, more books will be stacked to act as guides so that things do not topple, yet they do not support the weights and the syringe must be readable.
8) The first weight that will be placed on the syringe end cap plunger will probably need its hole covered. A thin yet durable piece of cardboard can be taped on the weight for this. If significant in mass, be sure to reweigh the weight for the first reading. If not strong enough, use a thin board instead.

## Procedure :

1) In doing the activity, be sure to have parental permission, assistance, and supervision. Always employ safety – such as wearing goggles. Parents should do parts of the activity where they deem necessary. Always exercise caution.
2) Draw back the plunger on the syringe to 50 mL and then secure it in the prepared block of wood.
3) If possible develop a clamp for the syringe ( that still allows reading access ) but in any case maintain balance of the system.
4) Now in sequence place the weights on the plunger.
5) Record both the weight total and the reading on the syringe volume. ( For example 2 one kilogram masses means 2 kg of mass, and 19.6 N of force for that volume ).
6) Note that all weights will be calculated in Newtons, while Pressure will be in Pascals ( which will be the weight divided by the area of the plunger in square meters ).
7) Record as many weights and volumes as you have for your data set.
8) Perform all the necessary calculations for Pressure.
9) Next calculate the product of a given Pressure and the Volume recorded to examine how close these products are to each other. Take the average of the products and determine the maximum error by the largest product difference from the average. It should be very small.
10) Graph Volume vs. Pressure ( it should be close to an inverse function $V = \frac{k}{P}$ )
11) Create a separate table of the log values of both Volume and Pressure. Graph these values on a linear graph ( or conversely graph the actual values on log scale graph paper ). Determine the slope of this graph. [ Note if done correctly it should be -1 ( don't concern yourself with the intercept ) ]

## Data :

Syringe Plunger Radius ( r )[cm] : _____

Syringe Surface Area ( A )[cm$^3$] : _____

| Trial | Weight ( F ) [N] | Pressure ( P=F/A ) Calculation [N/cm$^2$] | Volume (V) [cm$^3$] |
|---|---|---|---|
| 1 | | | |
| 2 | | | |

## Calculations :

Be sure to use a Slide Rule !

Formulae that may be needed :
Area of Circle : $A = \pi r^2$
Force due to Weight : $F = m*g$ ( take g = 9.8 m/s$^2$ )
Pressure : $P = F/A$
Note : 1 mL = 1 cc = 1 cm$^3$

Calculation Procedure : Plotting the Data :

1) From the data table convert the Pressure from **N/cm$^2$** to **N/m$^2$** and convert the volume from **cm$^3$** to **m$^3$**.
2) Create a table of the product of these values for comparison. ( P*V )
3) Graph on a linear graph Volume versus Pressure. It should not be linear and should be nearly an inverse function graph.
4) Look up using the Slide Rule the Log values of the Pressure and Volume and create a new table of these.
5) Graph the Log Volume versus Log Pressure. Draw a best fit line and determine the slope of this line.

## Conclusion :

Look at both of the graphs and the table of the products of Pressure and Volume. Your results should show that there is an inverse-relation between Pressure and Volume and their product should be nearly constant. This is verification through this process of Boyle's Law.

# Activity #15
## Pop Corn Pressure & Other Measures Activity
Grade Level : High School
Math Level : Challenging


One of the fascinating things is popping popcorn. Even here there is science at work. Corn is an essential food item and was a very important staple of the native Americans. Today it is on the cob, off the cob in many dishes, and we even use corn oil in cooking. Also there is an effort to employ corn in the production of fuel for automobiles.

Popcorn is a common snack, particularly in front of the TV or at movies. It requires the heating of the corn kernels obviously. But there is more to this. The reason it pops is the water vapor in the kernel. In fact, if there is a sufficient hole in the popcorn kernel, it will not pop.

The heat comes from some source and passes through the pan the kernels are in by the process of conduction. The heat then flows primarily via convection through the oil acting as a medium used to the kernels. The oil also serves to soften the shell slightly. The heated water in the kernel reaches its boiling point, causing the kernel casing to burst, turning the kernel inside out. This in turn releases the trapped water vapor. The carbohydrates inside now turn into the fluffy white popcorn we are looking forward to.

In this Activity, with parental permission and supervision ( be sure to read the safety rules and notes ) we examine the pressure of the bursting popcorn by measuring it indirectly and making some assumptions. We will make use of the Ideal Gas Law ( $P*V = n*R*T$ ) to find this pressure. For more on the Ideal Gas Law concept read Charles' Law Activity.

On the side, we can do some other calculations, such as the average mass per kernel, the percent water vapor in the kernels, and the number of moles of water in the popcorn too.

Who knew a bowl of popcorn could have so much science in it?
Enjoy.

**Purpose :** To determine the percent of water in popcorn.

**Purpose :** To determine the moles of water in popcorn.

**Purpose :** To determine the pressure necessary to make a kernel of popcorn pop.

## Materials :

- Popcorn Kernels ( 15-20 in number ),
- Stove,
- Small Pan,
- Cooking Oil,
- Mass Scale,
- Graduated Cylinder,
- Aluminum Foil,
- Distilled Water,
- Paper Towels,
- Tongs,
- Goggles,
- Slide Rule

NOTE : In place of pan, foil, and oil one can choose to use a Hot Air Popper instead. If so, you must then have access to a Digital Multimeter that has a Thermocouple which can operate in a range of $0°C$ to $1000°C$. Your temperature reading will only be around $200°C$ to $230°C$ for a typical Hot Air Popper. Be sure to note alternate directions for this measurement!

**Safety Notes : ( Be sure to read before any Activity )**

1. First and foremost – you must have parent permission and supervision for this Activity. Your parents should not only monitor but also do the process of operating the stove.
2. Also in the case of the popcorn, though often considered a food, here this is a lab and hence is not eaten. With the completion of the lab, it should be discarded. You never eat items used in any lab Activity.
3. Be sure to have on your goggles at all times.
4. In any use of flame such as here, exercise great care and caution. Do not touch or handle hot objects. Do not get close to any flame. Watch your clothing and hair when it comes to flame – roll up sleeves, keep loose garments away from heat, tie back hair and keep it away from flame. Be sure to act safely in the use of the stove. Be sure to let objects cool for some time before any thoughts of handling to clean. Always have a safety plan in place to deal with any emergencies.
5. Watch that the popcorn does not burn.

## Procedure :

1. Be sure to read the safety rules and employ these. Wear your goggles, have parental permission and supervision.
2. Count out the number of popcorn kernels to be used ( 15 to 20 are all that is needed ). Note that the number of pieces of popcorn will be directly related to the amount of water lost in the process of popping – hence the smaller the number the more sensitive your mass scale needs to be. So it may be best to increase the number if an initial trial fails to yield measureable results. Record this value in the data table ( N ).
3. Measure the total mass of the popcorn pieces by placing them on a zeroed out mass scale with a piece of wax paper or paper towel on it and record this value ( $m_T$ ).
4. Fill a graduated cylinder sufficiently so that when the popcorn is placed in it all of the pieces will be submerged ( first try with the popcorn in the graduated cylinder and then remove before filling ). Record this initial value of water in the graduated cylinder ( $V_i$ ).
5. Put the popcorn in the water in the graduated cylinder and record the final volume ( $V_f$ ) reading of the water level in the graduated cylinder.
6. Empty the water – be sure to not to lose any kernels – then dump the kernels on some paper towel and dry them.
7. Use of a Pan, Oil & Foil Method for Popping Corn :
8. Use a sufficient size piece of aluminum foil to loosely cover the small pan to be used to pop the popcorn. Poke some holes in it with a fork or knife carefully. Also when making this 'lid' be sure to have one or two large enough holes to see through at a distance ( do not go up to it ever ) so as to be able to watch the popcorn and prevent it from burning.
9. Put a small oil in the pan to have a very shallow depth. Place the pan on the scale. Put the popcorn kernels in it and cover with the foil 'lid'.
10. Record this mass value for all of the items ( $M_u$ ).
11. Take the covered pan with the oil and popcorn to the stove and turn on the stove. Be sure to follow safety rules in operating the stove.
12. Be sure to have parents do the following : Only gently heat the pan. Move it back and forth gently.
13. Once the popcorn starts popping, listen carefully and shut off the stove when it is near the end, still gently shaking the pan and let the remaining pieces pop. Do not let the popcorn burn.
14. Set the pan aside, still covered and let it cool. This may take several minutes of time ( 10 or more is recommended ).
15. Now again place the pan, oil, and popped popcorn on the mass scale and record the mass value of the system ( $M_p$ ).
16. At this point, you can clean up the tools used and dispose of the popcorn and oil items in use.
17. Alternate Method of Popping Corn with a Hot Air Popper :
18. Note that this requires a Digital Multimeter with a Thermocouple that operates up to 1000°C.

19. As noted in the prior steps, you have measured the mass of the popcorn – this will be the Unpopped Mass ( $M_u$ ). At the end of the popping in this method, you must take all of the pieces and once again measure their mass ( $M_p$ ). The Difference of these is the amount of Water in the Popcorn ( $M_w$ ).

20. In operating the Hot Air Popper and measuring its temperature, first, it is best to let a parent do this measurement. Next, it is best to measure its operating temperature with no popcorn in it. So plug it in and turn it on. Use the thermocouple to measure the temperature of the system as far down as possible and allow it to reach maximum temperature ( T ) which will be used in the formula for Pressure. The Thermocouple can just touch the metal but be sure not to let the probe catch on anything.

21. Calculations :

22. As an aside determine the average mass of each kernel [ $m_{ave}$ ].

23. Determine the Volume of the Popcorn pieces [ $\Delta V$ which is V later on ].

24. Determine the Mass of the Water given off by the Popcorn [ $M_w$ ].

25. As an aside determine the percentage of the mass of water in the popcorn [ %W ].

26. Using the molar mass of water, determine the number of moles [ n ] released.

27. Assume that the oil's temperature was 225°C and use this as the gas temperature, using the determined volume of the popcorn [ $\Delta V$ ] as the volume [ V ], calculate the Pressure [ P ] of the gas ( the steam ) using the Ideal Gas Law.

28. With the Pressure calculated, determine the ratio of it to the Atmospheric Pressure.

29. From the conclusion, compare the pressure determined to the assumed normal atmospheric pressure.

## Data :

Number of Kernels Used [ N ] : _____

Total Mass of Kernels [ $m_T$ ] : _____ g

Volume of Kernels Determination Data :

        Final Volume of Water & Kernels [ Vf ] : _____ mL
        Initial Volume of Water [ Vi ] : _____ mL
        Volume of Kernels [ $\Delta V$ ] : _____ mL
    o  Hint : Note that $\Delta V$ will be V in a later formula

Mass of Water Loss Determination Data :

        Mass of Pan & Oil & Unpopped Kernels [ $M_u$ ] : _____ g
        Mass of Pan & Oil & Popped Kernels [ $M_p$ ] : _____ g
        Mass of Water Loss [ $M_w$ ] : _____ g

## Calculations :

Be sure to use your Slide Rule!

Volume of popcorn through Displacement Method Formula :

$$\Delta V = V_f - V_i$$

Mass of Water Lost in Popcorn Formula :

$$M_w = M_u - M_p$$

( $M_w$ = Mass of Water Lost, $M_u$ = Mass of unpopped popcorn & oil & pan, $M_p$ = Mass of popped popcorn & oil & pan )

Average Mass of a Kernel of Popcorn :

$$m_{ave} = \frac{\text{total mass of popcorn } [ m_T]}{\text{number of kernels } [ N ]}$$

Percent Water in Kernel of Popcorn :

$$\%W = \frac{\text{Total Mass of Water}}{\text{Total Mass of Popcorn Kernels}} \times 100\%$$

Number of Moles :

$$\#n = \frac{\text{\# grams of substance present}}{\text{gram molecular weight of substance}}$$

Ideal Gas Formula :

$$P*V = n*R*T$$

Re-arranged Ideal Gas Formula solving for Pressure :

$$P = \frac{n*R*T}{V}$$

<u>Constants & Conversions to Use :</u>

1 L = 1,000 mL

1 kg = 1,000 g

1 ATM = 760 torr = 760 mm Hg

R = 0.0821 l*ATM/mol*K

K = °C + 273

<u>Assumed Constants :</u>

Atmospheric Pressure ( unless you have a direct reading use this ) :
1 ATM = 14.7 lbs/sq.in. = 101.3 kPa = 760 torr

Boiling Point of common cooking oil : 225°C

Molecular Weight of Water $H_2O$ : 18.0 g/mole

## Conclusion :

What was the percentage of water in the popcorn? Do you think this is a large or small value? How do you think this affects the ability of the popcorn to pop? How does your calculated Pressure compare & contrast to everyday atmospheric pressures? What are the potential sources of error in this experiment and what could be done to correct these?

# Activity #16
## Heat of Fusion of Water Determination Activity
Grade Level : High School
Math Level : Challenging


In volume I of The Inquisitive Pioneer we explored the Specific Heat of Water and Specific Heat of a Metal in Activities where we found the amount of energy taken in or given off by a substance while it is in a given state or phase, but this generates a next question ( which will be explored here in this book in this Activity ) : What about a given substance as it transitions from one state to another? Do the same rules apply?

In some ways the rules of science do apply but are modified to reflect the phenomenon of going from one state to another. In both cases the amount of the substance, or mass ( m ) is important and relevant to the equation in question.

Unlike the specific heat relation, however, there is no change of temperature during this process. If one were to graph the temperature change of ice per unit time on a graph one would notice that the slope is positive, but there are some areas of zero slope! Once the ice reaches $0°C$ the slope becomes zero – that is the ice stays the same temperature for a given time duration ( dependent on the amount of mass ) and then returns to a positive slope once all water in liquid form initially at $0°C$.

Interestingly the same would be true for water that is continually cooled from a warm temperature and turned into ice. This same zero slope region on the graph would occur. It too is dependent on the amount of mass.

Realize, however, that in the former case of ice to liquid heat is still being applied during this time duration. Energy cannot be created nor destroyed yet is not being detected by the thermometer. Where is it and what is it doing?

This energy in the case of the ice is being absorbed and used to rearrange the dihydrogen oxide molecules from their lattice arrangement of a solid into the fluid form of a liquid.

Also logically recognize that the same amount of energy must be given off by the liquid water and used to rearrange its molecules into the lattice arrangement when it is cooling and turning into a solid.

This amount of energy has been measured and given a name ( like many characteristics of matter ) in science. It is referred to as the latent heat of fusion or just heat of fusion of a substance. For a given substance it has its own unique value. In the case of water it ( $L_{fw}$ ) is about 80 cal/g ( rounded from 79.6 ). This is called the latent heat ( $L_f$ ).

It is calculated with the following relation :

$$Q = mL$$

Though called the heat of fusion, we would often think that fusion means going from a liquid to solid – that is, to fuse. Note that each direction of fusing or melting are the same and only relate to the direction of the energy flow.

The goal of this Activity is to see if we can determine the latent heat of fusion for water and compare it to the accepted value.

Important to note is that this idea also happens again in the case of liquid being turned into steam ( boiling ) or going from steam to liquid ( condensation ). So with a pot of water being heated, the liquid reaches the boiling point of 100°C until it turns into steam it will stay at the same temperature and requires energy to break the bonds between the adjacent water molecules to vaporize them. This value is not the same as the heat of fusion and has its own unique value. It is called the latent heat of vaporization and for water is 540 cal/g (2256 kJ/kg). This means that each gram of steam at 100°C has an additional 540 cal than one gram of water at the same temperature. Note that each is far too dangerous to be in contact with. Even more interesting is that the boiling water will remain at 100°C until all gone regardless of the heat source temperature.

Both of these ideas are critical in understanding the nature of materials and their behaviors at various temperatures. This is especially true in the case of why steam is so much more dangerous than the boiling water ( both of which are dangerous – so do not touch or put yourself near it or in contact with it ).

**Purpose :** To use ice and hot water with mass and temperature measurements to determine the latent heat of fusion of water

## Materials :

- Water,
- Tea kettle,
- Stove,
- Lab-quality Thermometers ( 2 ),
- Pyrex Measuring Cup ( can be done with a 2 cup one, but 4 cup may be a better choice depending on amount of ice needed ),
- Ice Cubes,
- Mass Scale ( a 2kg kitchen weight scale should be good ),
- Small Thermos ( typically used for coffee ),
- Aluminum Foil,
- Straw or Stirring Stick,
- Slide Rule

Note : Be sure to have parental permission and supervision when handling things like hot water. It may be best for them to do the monitoring and pouring of it when needed.

Note : In place of the Thermos, you can carefully use two Styrofoam Cups nestled inside each other ( it is going to be our Calorimeter ) and you can use a lid for it or use the aluminum foil that would have been used for the thermos as the lid here.

Note : If lacking a Mass Scale, you can assume that water has a density of 1 g/cc and be sure to measure the volume of the water and convert the volume of the water into an assumed mass ( Mass equals density times Volume $M = \rho * V$ ). In the case of the ice, use measured amounts of water to make the cubes – this is probably Tablespoon measurements.

## **Procedure :**

1. Be sure to read the Procedure first, create a plan of action and then do the following. Note that there is more than one way to measure the masses involved in this experiment and two alternate roads are noted. Choose the one that works best for you and your equipment available.

2. The term Calorimeter will be used through the Procedure and will refer to either your use of a small thermos ( best choice if available ) or two nested Styrofoam cups. Have it set up and set aside to be used.

3. Since it is best to do at least 3 trials of this experiment, have enough water in the tea kettle for each of the trials ( and since it is hot, it will require little to no additional heating from one trial to the next ).

4. Refer to the note above if you do not have a mass scale with regards to how to determine mass of the ice. In the case of the hot water you have a choice if you have a mass scale – either use the scale where you have measured the mass of the empty measuring cup which is then subtracted from the mass of the measuring cup with the chosen amount of hot water to be used OR using the volume and mass relation noted in the Note above, estimate the amount of mass being used.

5. Note : You need to use enough hot water to melt the ice and of course, you need a large enough Calorimeter to hold that amount of water – so be sure to scale your materials appropriately. A good guess is a regular ice cube will need a little more than one cup of hot water.

6. Now you are ready to conduct the first trial ( Note that you will do all of the following at least 3 times ) :

7. You can use only one Thermometer if the following is true : Your ice cubes as stationed in your freezer of your refrigerator and it has a reliable thermometer to obtain the initial temperature of the ice ( $T_i$ ).

8. If you do not have the conditions noted above, perhaps you have either no reliable thermometer in your freezer or you have a small cooler from which you are taking the ice. In either case, you need to place one of the thermometers in the ice's environment of containment and let it stabilize and then read and record the temperature of the ice ( $T_i$ ).

9. In the meantime you can fill the tea kettle so that it would make a large pot of tea with water and put it on so that it is heating. Before it boils, but it has heated for some time ( considering the amount of water being used ) now turn it off for the time.

10. Pour ( or have a parent do this ) about 1 to 2 cups of hot water into the Pyrex measuring cup. ( These have spouts and handles, so this makes the use easier ).

11. Either use the estimated mass ( noted above in the Notes ) or if you have a mass scale, measure the mass as noted above ( taking the difference of the water and measuring cup less the mass of the measuring cup ). Record this value ( $m_{hw}$ ).

12. Now pour this hot water into the Calorimeter ( thermos or Styrofoam cups ). Cover with the piece of aluminum foil, poke a hole and insert the thermometer.

13. Let the thermometer come to a stable temperature in a minute or two and record this value ( $T_{hw}$ ).

14. Take 1 or 2 ice cubes and either record their estimated mass ( see Notes above on how to do this ) or measure their mass on the mass scale and record the results ( $m_i$ ).

15. Put the measured ice mass into the Calorimeter and slowly mix with the straw ( through another hold poked in the foil ).

16. Watch as the temperature falls and stop when it reaches a value between 8°C -12°C. If the ice has melted before this, add more ice, but realize you have to amend your value for mass of the ice and the Total Mass of the system!

17. Interesting to note is that we do not have to determine the Total Mass of the system yet, and there is another way ( assuming you could not measure the mass of the ice by itself – read on ).

18. Once we are in the needed range of temperature then stop the process. Record this temperature – it is the final temperature of the process ( $T_f$ ).

19. Pour the Calorimeter contents into the Pyrex measuring cup again ( make sure yours is large enough ).

20. Thus far we can use these values to find all that is needed, but there is one more way to find the mass of the ice if we could not initially do it – read the following idea :

21. You have two choices here : Either use the total volume and volume-to-mass conversion factor for water to estimate the mass of the now cold water OR once again use the mass scale to measure the total mass of the cup and cold water and then subtract out the mass of the measuring cup to find the Total Mass of the system ( $M_T$ ).

22. Calculations :

23. Determine the heat lost by the hot water ( $Q_{Lhw}$ ).

24. Determine the heat gained by the cold water ( $Q_{Gcw}$ ) – be sure to use the correct mass in each of these calculations!

25. Find the difference of heat lost and heat gained – this difference is the heat used by the ice to convert from solid to liquid ( $Q_{ice}$ ).

26. Using the change of state heat value for ice ( $Q_{ice}$ ) and knowing the mass of the ice ( $m_i$ ) now find your estimated value for latent heat of fusion for water.

27. Do this experiment at least 3 times and average your results.

28. Now compare and calculate your final value for the latent heat of fusion of water to the accepted value and determine your percent error.

## Data :

Do the following table for 3 trials :

| Item | Measure |
|---|---|
| Mass of Hot water used ( $m_{hw}$ ) | |
| Mass of Ice ( $m_i$ ) | |
| Total Mass of system ( $M_T$ ) | |
| Mass of the Measuring Cup          ( $m_c$ ) | |
| Initial Temperature of Hot water  ( $T_{hw}$ ) | |
| Initial Temperature of Ice ( $T_i$ ) | |
| Final Temperature of System        ( $T_f$ ) | |
| | |

## Calculations :

Be sure to use your Slide Rule!

Total Energy in a closed system :

$$Q_T = \sum Q_i = Q_G + Q_L = 0$$

( T = Total, G = Gained, L = Lost : Note - for gain, + for loss )
Rearranged Formula to be used here :

$$Q_{ice} = Q_{Lhw} - Q_{Gcw}$$
( hw = hot water, cw = cold water ( ice here ))

Heat Gained / Lost Formula for a substance in a given state :

$$Q = m*c*\Delta T$$
( m = mass, c = specific heat capacity of substance, $\Box$ = change of temperature of substance )

Heat Loss by Hot Water :

$$Q_{Lhw} = m_{hw}*c*( T_{hw} - T_f )$$

Heat Gained by Ice ( treated as water at $0°$ C – so use specific heat of water )

$$Q_{Gcw} = m_i*c*( T_f - T_i )$$

192

Heat Needed / Emitted by a substance changing state :

$$Q = m*L_f$$
( m = mass, $L_f$ = heat of fusion for a given substance )

$$Q_{ice} = m_i*L_f$$

Average Value Calculation :

$$X_{ave} = \frac{\sum_{i=1}^{n} x_i}{n}$$

Percent Error :

$$\%E = \frac{[\text{ Accepted Value-Calculated Value }]}{\text{Accepted Value}} * 100\%$$

Needed Conversions / Constants :
1 fl oz = 29.6 mL = 0.125 cups = 2 tbsp
1 oz = 28.3 g
1 kg = 2.2 lbs
1 lb = 16 oz
1 cal = 4.184 J
$T°C = (T°F - 32°)/1.8$
$\rho_w = 1$ g/cc
1 cc = 1 mL
$c_w = 1.0$ cal/(g*°C)
$L_{fw} = 80$ cal/g = 333 kJ/kg

## Conclusion :
How close were each of your results to the accepted value for the latent heat of water? Did averaging the values bring the value any closer to the accepted value? In each case, why do you think this did or did not work well?

## Activity #17
## Freezing Point Depression Constant Activity
Comparison of the Freezing Point of Water & Salt- Water Solutions Activity
Grade Level : High School
Math Level : Challenging

We have read of the boiling points and freezing points of substances, such as water, but then this begs the question : Are there things that can affect these points? In particular we will explore the common phenomenon of adding salt to ice to assist in its removal during the winter.

With water on the ground, as temperature decreases, the water molecules slow down ( kinetic theory of matter ). At the right point of temperature and pressure, the molecules have slowed to a point so that the intermolecule attractions are too strong. Water molecules form their familiar lattice structures which we see as ice crystals.

As in all phase or state changes there are molecules entering and leaving the newly forming solid. When these two events are in equilibrium, this temperature point is the freezing point of that substance. For water we see it as $0°$ C or $32°$ F.

When there is salt present however, it act as a blocker and disrupts the equilibrium of the process. Salt prevent water molecules from entering the solid while not preventing the ones leaving the ice – essentially allowing the solid ice to return to a liquid state at its normal freezing point.

As temperature drops with salt present, the water molecules continue to slow further and the rates of water molecules entering the solid will match the ice particles leaving the solid. When these are now in equilibrium then this new temperature is the new freezing point given these conditions.

The effect of different materials, like salt, on ice formation is also called freezing point depression. There is a relationship of how much depression should occur for a given amount of water and salt concentration.

It is important to note that things like salt only have an effective temperature range over which it will work. When too cold, it will not have the effects wanted.

Our Activity will explore different salt-water solution concentrations and the temperature at which ice crystal formation takes place. We will use our actual observations as the experimental or measured data and compare it to the formula and its expected outcome given the conditions of our lab. These latter values will be treated as the Actual values and a percent error can be determined. A good idea is to repeat the exercise for a given salt-water concentration and average these results to reach a closer to true value.

**Purpose :** To determine the molal freezing point depression and compare it to our experimental results.

## Materials :

- Freezer ( We will use this to create salt-water 'ice cubes' for our testing freezer for the salt-water solutions ),
- Water,
- Salt,
- Ice Cube Trays,
- Small Cooler,
- Small Plastic Cups – if possible the best choice is Test Tubes in place of the small cups,
- Measuring Cup with metric measures,
- Measuring Spoons,
- Mass Scale,
- Pitcher ( measuring in liters – 1 L is perfect ),
- Thermometers – as many as cups are used ( as many as 3 or 4 ),
- Tall Glasses to hold Salt-Water Solutions ( good idea to use plastic since it is best to label these with tape and marker or just a marker ), - Note can used cleaned out regular pop bottles,
- Slide Rule

Note : What if you have only 2 thermometers available. Then continually redo the experiment trying to keep all of the essential components the same ( aka controls ) only changing the key component ( the independent variable ) which is the variety of solution.

Note : If unable to obtain a measuring cup ( typically a glass one ) that has metric measures, then use conventional measuring cups with English measurements on them, but employ the table of conversion values provided so that you can create the correct concentration for our salt water solutions.

Note : What if you do not have a Mass Scale? The best that can be done in this case is to approximate you measures of mass using the table of conversions provided to try and approximate the needed amount of mass of table salt ( NaCl ). You will mainly then use measuring spoons in various combinations. This sort of estimation does affect the overall results. In that case it may be best to do the experiment several times for a given solution and average the results. In a very good scenario, do a given solution 5 times, cross out the highest and lowest values and use the middle three values as the ones to determine an average from.  ** Note that this is also true for the water in terms of using conversions though the estimates are probably easier to bring close to the needed values for the solutions.

## **Procedure :**

1. Always read through the Procedure first in order to create a mental map of what to do, in what order, and where there may be critical points of focus.
2. After gathering your materials the first thing to do is to create the Salt-Water Cubes-slurry and Solutions needed for the Activity. In your materials be sure to look at the Notes section after the Materials list since these give you some alternatives. The Procedure will be written in the assumption you have a means to determine mass of the salt ( either directly with a scale ) or indirectly ( with measuring spoons ) and merely refer to it as the needed mass for a given solution.
3. 
4. Creating the Slat-Water Ice Cubes-slurry for the Activity Freezer :
5. 
6. Our first goal is to create salt-water ice cubes that will be put into the small cooler and used as a freezer chamber for the test tubes ( or small plastic cups ) filled with the various salt-water solutions of differing concentrations.
7. Note that this may take a day or a bit longer since we must create the ice cubes.
8. It is best to use 1 L of water and add to it 200 g of dissolved table salt ( NaCl ). This mixture is poured into our ice trays, placed in the freezer and left overnight. They will not freeze – but they will be colder than conventional ice and hence this mixture of ice and salt-water will be used in our testing freezer ( the small cooler ).
9. If there is no 1 L pitcher available, use the measuring cup if it has metric measures on it and fill a pitcher to 1000 mL. If no metric measures then use the conversion table below for converting English measures to metric ones to have the best approximation.
10. In the case of no mass scale, use the conversions below to obtain the 200 g needed.
11. 
12. Creating the Salt-Water Solutions :
13. 
14. In each cup measure 100 mL of water for the number of salt-water solutions to be tested. A good number is 4 and they will be of the following concentrations : 0.5 M, 1.0 M, 1.5 M, and 2.0 M
15. For the 0.5 M solution add 2.90 g of table salt,
16. For the 1.0 M solution add 5.80 g of table salt,
17. For the 1.5 M solution add 8.70 g of table salt,
18. For the 2.0 M solution add 11.6 g of table salt,
19. Be sure to label each of the salt-water solutions.
20. Also note that if you do not have 4 thermometers, it may be a good idea to instead run 4 trials where each trial is the same concentration of a solution. For example if you have 2 thermometers, do 4 trials, the first is 0.5 M, and the second will be the 1.0 M and so on – be sure to have enough ice cubes for this – you may have to extend the time frame for the Activity to several days.
21. Even in the case of having 4 thermometers and you are able to run all of the salt-water solutions at a time, it may be a good idea to consider redoing the Activity with a second trial to see if the results are similar or not. Also this enables you to average out the results.
22.

23. Freezing Point Depression Determination Activity Day :
24.
25. First put all of your ice slurry partially frozen salt-water ice cubes into the small cooler you have available. Make the depth so that the test tubes or cups being used can sit in them and not tip over or be affected by this mixture. If needs be add some conventional ice from a bag of ice. This cooler will be called our testing freezer.
26. It is best to use test tubes, but you can use small plastic cups – in fact ones with a color are best because you have to observe them as the solution in them starts to freeze and form ice crystals. In either case use a small enough measuring cup so that when full with a given solution and then poured into the testing vessel ( tube or cup ) it does not overflow and is about ½ to ¾ full at most.
27. It is best to label the testing vessels so that you know which of your solutions goes into that vessel
28. Fill each of the testing vessels with each of the solutions – Note It is important to rinse the measuring cup between each of the solutions – you do not want cross-contamination – so rinse the measuring cup with water and dry it between each of the fillings.
29. Carefully place each of the testing vessels into the testing freezer. Put a thermometer into each of the testing vessels.
30. Regularly watch each of the testing vessels. You are looking for ice crystal formation on the surface of the liquid. Once the crystals begin to form, check the temperature of the solution in question and record this on the Table provided. Once all solutions in the Trial are done and have recorded values you can either do another Trial run or move onto the Calculations.
31.
32. Calculations :
33.
34. First, if your thermometers have recorded the temperature in Fahrenheit then they must be converted into Celsius – use the formula in the calculations section of the Activity.
35. If you have performed multiple Trials for a given solution, then use the data to calculate an Average Temperature using the Average formula.
36. Each of these determined Temperatures is our Change of Temperature as found in the Activity. They will be compared to the Predicted Temperature Change from the known formula.
37. It is a good idea to graph the Temperature of Freezing Point vs. Molal Mass of Solution. Each is its own data point and it is best to have a bar graph of these. Ask yourself if there are any trends in the data?
38. Determining the Predicted Change of Temperature
39. Though the table above in the Procedure helps you set up the correct molality of the NaCl in the water solution, we can, for practice determine the molality ourselves with the following steps :

40. First sum up the atomic mass of sodium and chlorine in grams to find the molecular weight of the molecule in grams for one mole of the substance. This value is the denominator of the # moles equation. The number of grams used in each case for the solution of salt is the value to place in the numerator. The determination of this ratio is a number and is the number of moles of sodium chloride used for each solution. ( Note : you will have to do this step if you decided on your own values for mass of salt used ).

41. We are assuming that the density of water is constant at 1 g/cc, but in reality it does depend on its temperature – it can range from 0.996 to 1.0 in our Activity. Using 1 g/cc as our standard, we also use that 1 mL of water will have a mass of 1 g and we find that each of your solutions have 0.100 kg of water in them.

42. We can now determine the molality of each of the solutions. We take the number of moles divided by the mass of water ( in kg ) for each of the solutions.

43. We can now use the Change of Temperature Freezing Point Depression Formula ( $\Delta T$ ) to find our predicted value of temperature depression and compare it to our results.

44. You can now compare these Predicted values to our Actual values measured in the Activity. We will assume that the Predicted values are the Accepted Values and the Actual values from the Activity are the Measured Values and then employ the Percent Error formula to find percent error for each of these.

## Data :

Use the following Data Table, but if doing more than 1 trial per solution create the needed number of Tables where each is a Trial.

| Solution | Freezing Point of Solution ( °C ) |
|---|---|
| Pure Water ( if using a control ) | |
| 0.5 M salt-water solution | |
| 1.0 M salt-water solution | |
| 1.5 M salt-water solution | |
| 2.0 M salt-water solution | |

## Calculations :

Be sure to use your Slide Rule!

Molal Freezing Point Depression Constant or Cryoscopic Constant ( $K_f$ )

$$K_f = \frac{\Delta T}{i*m}$$

$K_f$ for water is 1.86 $^\circ$C*kg/mol, this is the molal freezing point depression
constant or also known as the cryoscopic constant
i = van't Hoff Factor ( 2 for NaCl ),
m = molality of the solute in mol solute/kg solvent

Change of Temperature Freezing Point Depression Formula :

$$\Delta T = i*m*K_f$$

Molality ( m ) – The ratio of the number of moles of solute dissolved in one
kilogram of solvent

$$m = \frac{\# \text{ Moles}}{\text{Mass of solvent ( here water } M_w \text{ ) ( kg )}}$$

Number of Moles :

$$\# \text{ Moles} = \frac{\text{Given Mass in question}}{\text{Gram Molecular Weight of Molecule in Question}}$$

Gram Molecular Weight of a Molecule = Molar Mass expressed in g/mol =
Is the sum of all the molar masses for each of the components of the
compound in question

Mass of Water : ( Use formula if you do not have a mass scale and need the
mass of the water and this is estimated from its volume )

$M_w = \rho * V$
( Mass equals density times volume )

Average Value Calculation :

$$X_{ave} = \frac{\sum_{i=1}^{n} x_i}{n}$$

Percent Error :

$$\%E = \frac{[\text{ Accepted Value-Calculated Value }]}{\text{Accepted Value}} * 100\%$$

Table of Needed Conversions / Constants  ( rounded to 3 sig figs ) :

$T^\circ C = (T^\circ F - 32^\circ)/1.8$
1 fl oz = 29.6 mL = 0.125 cups = 2 tbsp
1 oz = 28.3 g
1 kg = 2.2 lbs = 1000g
1 lb = 16 oz
$\rho_w$ = 1 g/cc
1 cc = 1 mL
1 L = 33.8 oz = 4.22 cups = 1.06 quarts
Table Salt ( NaCl ) Mass Approximations :
      1 Tbsp = 18.3 g
      1 tsp = 6.10 g
      ½ tsp = 3.05 g
      ¼ tsp = 1.52 g
      1 oz = 4.98 tsps
Atomic Weight of Sodium 35.5 g in one mole
Atomic Weight of Chlorine 23.0 g in one mole
i = the van't Hoff factor – a constant associated with the
  amount of dissociation of the solute in the solvent. For
  substances that do not dissociate, i=1. Here in our
  experiment in using salt ( solute ) and water ( solvent ),
  the salt ( NaCl ) will completely dissociate into two ions
  ( Na+, Cl- ) so here i=2

## **Conclusion :**
What do your results show as you increased the concentration of salt in the solution? How does a lower freezing point make salt effective as an agent in de-icing the roads in the winter? Do you think there are temperatures it will not work at or concentrations that will not return the same level of results?

# Activity # 18
## Pre- & Post- 1982 Penny Density Determination Activity
Grade Level : Middle School
Math Level : Calculating


With regards to the concept of Density, for more information look to volume I of The Inquisitive Pioneer for other density activities as well as the web site : www.cosmicquestthinker.com . Essentially density is a physical property of matter and is the ratio of mass to volume for a given material.

Note that it is an average for the whole object when looked at collectively and the density of any part or portion of a substance much vary greatly from the average. Take, for example, a boat made of aluminum. Aluminum itself is much more dense than water ( 2.7 x ) but the total mass of the boat divided by the volume it occupies will yield a value less than the density of water ( 1 g/cc ) hence the reason why it floats!

If one can determine the average density of a material, it can be compared to another object. In this Activity we are going to compare one penny to another. Though this seems mundane and obvious, it will depend on when the penny was made. US pennies before 1982 were 75% copper and 25% zinc in composition, while those post 1982 where 97.5% zinc and only 2.5% copper ( due to commodity versus production costs ). So in our Activity we will compare pre and post 1982 pennies. Note, it may take some time to find these.

Note : If unable to find the requested 25 of these, then use what you have and adjust the tables accordingly to your outcome ( for example, let's say you find 16, then have the table read : 4, 8, 12, 16 ).

Much like the aforementioned note on the aluminum boat, we can take the information presented here and put this idea to use. We know the percentage of each of the pennies composition and with tabular information on the density of copper and zinc we can then compute mathematically the expected density of these pre and post 1982 pennies. These computed values can then be compared to our actual experimental results.

This Activity serves to illustrate one of the most important aspects of density values. They tend to be averages for a given total material and do not indicate the particular density of a given component of a material. This type of information can then be determined through careful investigation if needed. Also, the average serves to give an idea of what a composite material ( an item made up of more than one type of substance ) may be similar to.

**Purpose :** To determine and compare the density ( from mass and volumetric displacement means ) of US pennies made pre-1982 ( 75% Cu, 25 % Zn ) to US pennies made post-1982 ( 97.5% Zn, 2.5% Cu )

## Materials :

- 25 pennies pre 1982,
- 25 pennies post 1982,
- 100 mL graduated cylinder ( large enough to allow pennies to be put in ),
- Mass Scale,
- Small Cup,
- Water,
- Graph Paper,
- Slide Rule

## Procedure :

1. Use the cup to put water into the Graduated Cylinder. It is best to start with about 40-50 mL. Whatever the value, read and record this value in your table in Trial 1 in the initial value.
2. Be sure to have two sets of pennies : the Pre-1982 set and the Post-1982 set. Note that for each, the data tables in the Data section are to be used.
3. Using the Mass Scale measure the mass of 5 pennies in the set together and record this value ( note : the number of pennies used depends on how many you have an how many you have in each of your trials – see above Note ).
4. With each set of 5 pennies, you have to write down the mass for the given set of 5 and then add it to the ever-growing total mass. For example in the second set of pennies, the mass of these 5 pennies is added to the mass of the first 5 pennies and this total is placed in the Total Mass column.
5. Add the set of pennies, one at a time and carefully, to the Graduated Cylinder. Once all of the set of 5 are in, record the Final Volume for the given trial.
6. Note that the Final Volume for a given trial is now the Initial Volume for the next trial and the next set of 5 pennies!
7. Repeat the process of measuring the mass of 5 pennies at a time and placing them in the ever-growing collection of pennies in the graduated cylinder and recording the final volume after each set is added.
8. Complete both sets of pennies ( pre-1982 and post-1982 ).
9. Calculations :
10. For each set of data ( pre-1982 and post-1982 pennies do the following steps :
11. First determine the Change in Volume which is the difference of the Final Volume and the Initial Volume ( $\Delta V = V_f - V_i$ )

12. In each step of the Table determine the Total Volume for that set of pennies by summing up the Change in Volume from that trial and all prior trials. ( For example, the total volume for 10 pennies is the Change in Volume for Trial 2 added to the Change in Volume in Trial 1 ).
13. Graph the Total Volume vs. Mass on a x-y axis ( Total Volume is the x-axis and Mass is the y-axis ). ( For example, the mass of 5 pennies is the y-variable while the volume for 5 pennies is the x-variable, and so on ).
14. Draw a best fit line through the points on the graph ( to the origin ) and determine the slope of the line from two points on the graph.
15. Calculate the Density of each of the trials where the mass for those 5 pennies in a given trial is divided by the change of volume for that trial.
16. Sum up all of the density values you calculated and calculate an average density for this.
17. Compare your Average Density Calculation to your Slope. Are they similar? Should they be?
18. If you want to compare your results to what is expected, use the formula in the Calculations for Density from Percent Composition to determine the Density for both the pre-1982 and post-1982 pennies and compare these to your calculated values for both sets of data. Use the percent error formula to find how much variance you have.

## Data :

Each of the following tables is needed separately for the set of pre 1982 and post 1982 pennies.

| Trial | Number of Pennies | Total Mass ( g ) |
|---|---|---|
| 1 | 5 | |
| 2 | 10 | |
| 3 | 15 | |
| 4 | 20 | |
| 5 | 25 | |

Note : Read the Data Table carefully. It notes the Total Mass for the Set of Pennies. You need to write down the mass of each set of 5 pennies and add these and place the Total Mass in the column!

| Trial | Number of Pennies | Vf : Final Volume ( mL ) | Vi : Initial Volume ( mL ) | $\Delta V$ : Change in Volume ( mL ) | Total Volume of Pennies : $V_T$ ( mL ) |
|---|---|---|---|---|---|
| 1 | 5 | | | | |
| 2 | 10 | | | | |
| 3 | 15 | | | | |
| 4 | 20 | | | | |
| 5 | 25 | | | | |

## Calculations :

Be sure to use your Slide Rule!

Slope :

$$m = \frac{\Delta Y}{\Delta X}$$

Change :

$$\Delta V = V_f - V_i$$

( The difference in a variable is the difference of the final and initial values for that variable )

Total :

$$V_T = \Sigma\, V_X$$

( The total of a Variable is the Sum of the components or parts )

Density :

$$\rho = \frac{M}{V}$$

( M = mass, V = volume )

Computed Density from Percent Composition :

$$\rho = \Sigma\, \frac{A\%}{100} * \rho_X$$

The density is the sum of the percentages expressed as a decimal times the density of that given material. Here there are only two factors to consider : the copper and the zinc. Note 'x' refers to each given element.

Needed Information :
Density of Copper : 8.92 g/cc
Density of Zinc : 7.14 g/cc

Note : First use these values to find the Accepted Value for the density for pre and post 1982 pennies and then use them in the Percent Error formula

Percent Error :

$$\%E = \frac{[\text{ Measured Value-Accepted Value }]}{\text{Accepted Value}} *100\%$$

## **Conclusion :**

How similar are your density values to each other? How close are they to the expected values?

# Activity #19
## Floating Objects and Archimedes Principle Activity
Grade Level : Middle School
Math Level : Calculating

Archimedes History and Principle :

The idea of floating and sinking objects ( our next two Activities ) first became a concern for Archimedes from the concept we call density today. The density of a given geometric form is easily determined once there is a scale for the mass of the object in question and since the object has a regular geometric form, its volume can readily be calculated from mathematical formulae – density is mass per unit volume ( in this book we explore the density of various pennies from different years but this concept is further considered in volume one of The Inquisitive Pioneer ).

But what if the object's form is not regular? For example it is a lumpy stone. One could approximate the volume and with enough measurements may even find an upper and lower limit to it. This puzzle is one of the questions posed over 2000 years ago to Archimedes, one of the greatest of mathematicians and inventors of all time.

Archimedes of Syracuse ( c. 287 BC – c 212 BC ) is best described as a Greek mathematician, inventor, engineer, physicist, and astronomer. Some of his notable claims to fame are his explanation of the lever ( another of our Activities in this book ), the Archimedes Screw to act as a pump, a useful value for pi ( $\pi$ ) of 3.1416, an approximation for the $\sqrt{3}$ , and in his work, The Method, the basis of modern day calculus.

One of the more famous stories is the basis of this activity and can simply be referred to as the Golden Crown. King Hiero II had a crown made by a goldsmith whom he felt may have cheated him by withholding some of the supplied gold to make the laurel-leaf shaped crown. The problem is how can one determine the crown to be pure gold without melting it back down and then comparing that amount to the amount of gold used. It is said that while attending a bathhouse to bathe he noted the amount of water being displaced when entering the tub and recognized that the amount displaced is equal to the volume of the item placed in the tub. "Eureka" ( Greek, meaning "I have found it!" ) he shouts as he races sans clothing from the bathhouse.

The exact method of gold determination was probably based on his work in hydrostatics ( such as found in his work, On Floating Bodies ) and buoyant forces ( he gives us the Buoyant Force ). Here this method looks for a difference in buoyant force. The more water displaced the greater the buoyant force basically.

His idea of Buoyancy is the basis of why objects sink or float. A ship either made of wood or even metal must displace its weight in order to float. The weight displaced is based on the volume of fluid, here water, which must equal the weight of the vessel. Hence a floating object has no net force acting on it since the force due to weight is equal to the buoyant force of the fluid it is in. An object sinks when its displaced weight is less than its weight. So a block of metal in air weighs more than one under water. The one under water has a net weight equal to its weight in air less the buoyant force acting on it.

The concept of buoyancy applies to any mass acting in any fluid. Not only is water a fluid, but so is air. So this idea applies to explaining and analyzing why hot air balloons and helium-filled balloons float ( this book also has a helium-filled balloon activity too ).

Not only can this idea be examined from the perspective of buoyancy, but conclusions can be drawn about the density of the object and the fluid it is in. If the overall average density of the object, say a metal ship, is less than the fluid it is in, here water, then it floats. If it is greater than the fluid, then it sinks. If equal, then it has neutral buoyancy and can be placed anywhere, such as in the case of a submarine.

Also this is why people tend to float ( lower density than that of water ) as well as the fact that nearly all forms of wood float in water ( since they have density values less than water ).

Returning and Examining the Gold Crown story and the use of Buoyancy and Density :

First the crown is balanced on a scale beam with the crown at one end and the needed amount of gold at the other. This balanced system is then placed in the water. If the crown is equally dense the rod would stay straight since both displace the same volume. If the crown was more dense for some reason it would displace less volume hence a smaller buoyant force so the scale would tip in the direction of the crown. If, however, it was less dense, the crown would displace more water, hence have a greater buoyant force and the scale rod would tip in the direction of the gold mass at the other end.

The other oftentimes reported method noted the amount of displaced water by both the crown and the equal mass of gold. If a greater amount of water for the crown, means a greater volume, hence a lower density.

In either case, the investigation notices the property of density. It is reported that the crown was less dense hence had been blended with other less dense materials and the goldsmith had pocketed some of the gold.

The value of density cannot be overestimated in any case. The physical property of density is one amongst many to characterize a material to begin with. Next, it then can be used to quickly determine its ability to float or not and the level to which it will sit in a fluid when it floats. With the use of density by Archimedes, a major step in the direction of scientific analysis was taken in history.

Another Activity is in this book ( #19 ) which involves Archimedes Principle ( with regards to sinking objects totally immersed in water and density through water displacement )as well.

This Activity focuses on the concept of floating objects. To float an object must have an upward force to balance its weight force which points down. These forces must be balanced ( a net force of zero ) otherwise ( according to Newton's 1$^{st}$ Law ) the object would be accelerating up or down, but it is stationary ( one of two possible states when forces are balanced ).

This upward force is called the Buoyant Force and comes to us from Archimedes who noted this idea long ago. The weight of the displaced liquid when an object is placed in it is equal to the weight of the object. Hence a boat that weights 2 tonnes, even if it is made of metal, when it displaces 2 tonnes of water will float. The key in boats is to shape the material so that it spans a large volume so as to displace as much water as possible ( yet at the same time having a streamlined shape for mobility in the water, since water will act as a drag as the motor powers it through the water ).

Note that the volume of the water for the floating object is NOT equal to the volume of the boat itself! These values are only the same when the object is completely immersed in the water ( hence sunk or submerged ).

An interesting consequence of the fact that a floating object requires less volume than itself in the volume of displaced water means that water itself is more dense than the boat! How odd this must seem?! A metal boat having less density than that of water itself. Note that this is the average density and not the density of any particular piece of the boat. The density of the metal of the body does not change in the water – it remains what it is in air ( for example aluminum is 2.7 g/cc ). But when you take the total mass of the boat and divide it by its entire volume that it encompasses ( all the air space too ) then it turns out that its average density will be less than the density of water ( 1.0 g/cc ).

The goal of the Activity is to try various 'boats' of differing masses, to measure the amount of displaced water by volume and calculate its mass and compare this to the mass of the boat itself. The displaced mass should tell us the buoyant force acting on the 'boat'.

**<u>Purpose :</u>** Compare the Mass of Displaced Water by a Floating Object
Experimentally to the Calculated Mass of Expected Displaced Water by
a Floating Object using Archimedes Principle.

## <u>Materials :</u>

- 16 oz. Plastic Cups ( at least 2 ),
- Mass Scale,
- Marbles,
- Water,
- Marker,
- 4 c. Measuring Cup,
- Optional Equipment : Tension Scale, Graduated Cylinder,
- If a 4 c. Measuring Cup not available use a square pan and a smaller measuring cup ( 2 cup size ),
- Slide Rule

## <u>Pictures :</u>

## <u>Procedure :</u>

1. The overall cups system will be placed in a container to catch the displaced water ( either a square pan or a 4 cup measuring cup ). Place the cup for water in this container.
2. Fill this first cup up to the very top carefully.
3. In the other cup now decide to use a set of marbles ( 10 minimally for each activity – up to around 40 total ) in an incremental manner for each data set ( for example start with 10, then 20 in the next set, et al ). Place this chosen number of marbles in the second cup and take its mass ( B ).

4. Now slowly and carefully let the boat settle into the first cup which has water in it. The water should displace into another container ( see no. 1 ).
5. Measure the displaced volume of water ( ΔV ). Do this 3 times, recording the displaced volume each time.
6. Note : Measured displaced water in the measuring cup is easiest. In the absence of such an item you need to pour the water from the square pan into a measuring system ( another smaller measuring cup or the use of a graduated cylinder ).
7. Note : Be certain to not spill the water. Also be clear as to the degree of precision of the tools you are using when recording your results.
8. Do not quickly discard this water since you must measure its mass too for comparative purposes ( M ) and record it.
9. Reset the system and do the Activity at least 3 times ( 3 different sets of marbles making up the bulk of the mass in your boat ).
10. Calculations :
11. Though the Slide Rule is a recommended tool, all of these calculations can be done with a regular or scientific calculator. Some scientific ones even have built-in averaging formulae. For those who like spreadsheets, the data can be typed in and the formulae then also be typed in its own cell where the formula references each of the measured variables in their respective cells ( for example in cell B1 the mass of the boat ( B ) is typed and a later cell, say B3 has the formula where it occurs ).
12. Determine the Average ( $V_{ave}$ ) amount of Displaced Water for each 'Boat' in your Activity.
13. Using the Average Displaced Volume of Water, calculate the Weight of the Water Displaced ( $F_B$ ).
14. Convert both the mass of the 'Boat' ( B ) and the mass of the measured displaced water ( M ) into weight.
15. Compare all of the weight values ( they should ideally be identical ).


## Data :

Note : Recreate the Table for each 'Boat' you test & for each 'Boat' run
        3 trials and average the volume displaced :

Mass of the 'Boat' [ B ] : _____ ( kg )

| Trial | Measured Volume of Water Displaced ( ΔV ) |
|-------|-------------------------------------------|
| 1     |                                           |
| 2     |                                           |
| 3     |                                           |

Mass of Displaced Water for Comparative Purposes [ M ] : _____ ( kg )

## Calculations :

Be sure to use your Slide Rule!

**Average :**

$$V_{ave} = \frac{\sum_{i=1}^{n} V_i}{n}$$

**Calculated Weight of Displaced Water :**

$$F_B = D*g*V_{ave}$$

**Converting Mass to Weight :**

$$W = B*g$$

**Conversions & Constants that may be needed :**
Acceleration due to gravity for Earth :

$$g = 9.8 \text{ m/s}^2 = 980 \text{ cm/s}^2$$

Density of Water : $D_{water} = 1$ g/cc $= 1000$ kg/m$^3$

1 m = 100 cm
1 mL = 1cc = 1cm$^3$
1kg = 2.2 lbs.
1 N = $\frac{1 \text{ kg*m}}{s^2}$ = 0.224 lbs.
1 oz. = 28.3 g
1 oz. = 29.6 mL
1 qt. = 4 cups
1 cup = 8 ozs.
1L = 1000 mL = 1.06 qts.

## Conclusion :

As an alternative in measuring, use the marker to mark the side of the 'boat' cup when it is floating in the other cup of water and determine a way to measure this amount of displaced water up to this volume of the cup ( besides measuring the displaced water itself ).

Did your results show that the weight of the 'boat' equals the weight of the displaced water? What does this mean in terms of a net force acting on the 'boat'?

# Activity #20
## Archimedes Principle Activity pt II Objects that sink :
Grade Level : Middle School
Math Level : Calculating

In a short summary of Archimedes history, Archimedes found that the weight of an object in air and when placed in water are different. It is different by what we refer to as the Buoyant Force and this in turn is equal to the weight of the displaced water, which equals the volume of the object. He famously used this idea to solve the question of whether a king's crown was made of pure gold or not.

Using this idea, it is easy to find the volume of an irregularly shaped object and with a good mass scale, these two quantities, mass and volume, allows the determination of density of the object, which can be a useful physical property of a given material.

Further, Archimedes Principle allows for a simple conclusion to be reached : If the density of the object placed into any fluid ( liquid or gas ) is greater than the density of the surrounding fluid ( or medium ) it will sink. If it is less, then it will float. This is the reason why large steel-hulled ships float – they are less dense than the water they are in. They are able to displace their weight. As the ship is loaded, it sits further down in the water.

This is also the reason hot-air and helium-filled balloons float. They too are less dense than the surrounding air. In fact as long as there is a force acting on the rising balloon, the buoyant force in this case, the balloon will rise. When the forces of weight and buoyant force balance out, then the balloon stops rising. This is all in accord with Newton's Laws of Motion, where a Net Force results in a change of motion ( the rising of the balloon ) and when the forces balance it comes to rest, since now when rising further there is an exerted force downward ( the weight force ) and hence the balloon reaches equilibrium.

In this Activity, we directly and indirectly measure the Buoyant Force for comparative purposes as well as determine density for a given stone.

**Purpose :** To take measurements of mass, weight in air, weight in water, and water displaced to determine the buoyant force acting on a mass of density greater than that of water.

Note : There are several distinct purposes here :
1) Determining the Buoyant Force through Scale Measurements,
2) Calculate the Buoyant Force through Volume of water displaced and its Weight determination, and
3 ) Determination of the Density of an Object through use of the ideas in Archimedes Principle

## Materials :

- Small Bucket with Lip,
- Pan large enough for bucket to sit in and water over-spilling the bucket with immersed object will collect,
- Measuring Cup with markings ( metric and English best ),
- Note : If possible – use a Graduated Cylinder instead of measuring cup,
- Spring Mass/Weight Scale ( as used in fishing, pounds or kilograms ),
- Kitchen Mass Scale ( measurements in grams ),
- Water,
- Small palm-sized Stones,
- Dental Floss or Fishing Line String,
- Scissors,
- Slide Rule

## Procedure :

1) To determine if the bucket chosen is best, first try to put the stones to be used in it to see that they fit.
2) Next fill it to the top with water ( with the bucket in the collection pan for spilled out water ). Place a stone in. Be sure to have enough water so that the water spills out into the pan ( this can be tricky due to the surface tension of water ).
3) To set up all things – have the bucket in the pan and the bucket filled with water.
4) Have a small set of stones ( 3 – 6 ) all about palm-sized to be used.
5) First tie the string being used to the stone to be measured. ( Be careful with fish line as this can cut – have supervision or help ).
6) Measure the mass ( in grams ) of the stone on the kitchen scale and record this value ( M ).
7) Measure the weight of the stone ( in pounds ) with the spring scale in air and record this value ( $W_{air}$ ).
8) While the stone is on the string and on the spring scale, lower the stone into the water. Once underneath the water, measure and record the value of the weight reading on the spring scale ( $W_{water}$ ).
9) Be sure to have collected the water spilled out of the bucket when the stone was placed into the water. Carefully remove the stone, then the bucket and then pour the contents of water in the pan into the measuring cup and record this volume value ( $\Delta V$ ).
10) If possible, use a graduated cylinder to have more precise values of volume.
11) Perform all of these measurements aforementioned for each of the stones separately and place their values on a table of information.
12) Calculations to do for each of the stone's data :
13) First calculate the Buoyant Force ( $F_B$ ) ( in pounds ) directly from the difference of the measures of Weight in Air and the Weight in Water of the stone.
14) Convert the Volume measure ( $\Delta V$ ) from cups into $cm^3$ if using cups.
15) Determine the Density ( $D_{stone}$ ) of the Stone by using the ratio of the kitchen mass scale reading and the converted volume measure.

16) Calculate the Indirect Buoyant Force ( $F_{B2}$ ) from the displaced volume of water, the density of water, and the acceleration due to gravity.

17) Be sure to be using the proper units for the Force. Your initial calculation may have units of : $\frac{g*cm}{s^2}$ . this means you will have to convert the calculated buoyant force into Newtons and then Pounds. This will be the second buoyant force measure, called the Indirect Buoyant Force ( $F_{B2}$ ).

18) Compare this new predicted value to the originally directly determined Buoyant Force ( $F_B$ ).

19) Also for comparison, calculate the predicted weight in water of the stone using the indirect formula involving the use of densities of the stone and water.

20) Assuming that the direct measure is the correct value, what is the percent error of your calculated value and why would this occur? Can you find ways to eliminate the error involved?

21) Though the Slide Rule is a recommended tool, all of these calculations can be done with a regular or scientific calculator. Some scientific ones even have built-in averaging formulae. For those who like spreadsheets, the data can be typed in and the formulae then also be typed in its own cell where the formula references each of the measured variables in their respective cells, for example B1..BN has the measurements and values used in the equation while BN+1 has the formula for all of these variables ( why not A? Simple – use it to label you variables )

## Data :

Note : Give each stone an ID, such as a Number or Letter.

| Stone ID | $W_{air}$ (lbs) | $W_{water}$ (lbs) | $\Delta V$ (cups or mL ) | M (g) |
|---|---|---|---|---|
|  |  |  |  |  |
|  |  |  |  |  |
|  |  |  |  |  |

| Stone ID | Direct Buoyant Force $F_B$ (lbs) | Indirect Buoyant Force $F_{B2}$ (lbs) |
|---|---|---|
|  |  |  |
|  |  |  |
|  |  |  |

## Calculations :

Be sure to use your Slide Rule!

### Direct Determination of Buoyant Force formula :

$$F_B = W_{air} - W_{water}$$

### Indirect Determination of Weight in Water formula :

$$W_{Water} = g*\Delta V*( D_{stone} - D_{water} )$$

### Indirect Determination of Buoyant Force formula :

$$F_{B2} = g*\Delta V*D_{water}$$

### Density : Density is mass per unit volume

$$D = \frac{M}{V}$$

### Weight Force formula :

$$W = M*g$$

### Table of some values you may need :
Acceleration due to gravity for Earth :
$g = 9.8$ m/s$^2$ = 980 cm/s$^2$

Density of Water : $D_{water}$ = 1 g/cc = 1000 kg/m$^3$

1 m = 100 cm
1 mL = 1cc = 1cm$^3$
1kg = 2.2 lbs.
$1 N = \frac{1 kg*m}{s^2}$ = 0.224 lbs.
1 oz. = 28.3 g
1 oz. = 29.6 mL
1 qt. = 4 cups
1 cup = 8 ozs.
1L = 1000 mL = 1.06 qts.

### Conclusion :
The basic idea here is to compare and reflect on results and consider options to refine the measurements.

# Activity #21
## Buoyant Force & Charle's Law of a Helium Balloon at Various Temperatures Activity
Grade Level : Middle School
Math Level : Calculating

This Activity is 2 in 1 – one is Charles Law which should be done with a helium-filled balloon using cold temps ( such as the refrigerator or freezer ) while the other Activity examines the lifting capacity of a helium-filled balloon which should be done at warmer times of the year and use indoor, outdoor, and car environments for the balloon to test this idea.

The following examines the Charles Law aspects of the Activity :

Temperature is by definition a measure of the average kinetic energy of the particles making up a material. Kinetic energy is energy due to motion, so it stands to reason that as temperature increases, so too does the average energy of the particles in a material.

If the substance is a gas, particles of a lower temperature will move more slowly and bounce off each other less actively. On average, they will remain closer together than in the case of a higher temperature gas, where the particles will have greater speed and motion.

What is generally found is that as an object is heated, it expands and when cooled it contracts in volume.

This behavior of gases was first described by Joseph Louis Gay-Lussac in 1802 but credits it to Jacques Charles work from the 1780s.

Charles' Law is paraphrased as stating that when there is a constant pressure that the volume of a gas will increase or decrease by the same given factor as the temperature that the gas is at ( on an absolute temperature scale )

Charles' Law is also known as the Law of Volumes. As the temperature increases, the gas volume increases, and vice versa. Note that there are two critical ideas here minimally. First, this applies to an ideal gas, which is never the case, but many gases can exhibit this property to a certain degree. Also for all of these gases, the Pressure is held constant as well as the number of molecules under consideration. ( As an aside, the relation of Pressure and Volume is Boyle's Law and when are all combined it is the Ideal Gas Law where $\frac{P*V}{T}$ is a constant )

Like other ideal gas laws, this law is seen as an extension of the Ideal Gas Law and its equation itself. This equation derives from the Kinetic Theory of Gases that basically presumes that gases are composed of point-like spherically-shaped atoms/molecules that occupy negligible volume, do not attract each other, and undergo perfectly elastic collisions ( i.e. no loss of kinetic energy ).

216

An interesting consequence of this law is that at absolute zero, 0 K, a gas would occupy zero volume. This is impossible, since all gases would turn into liquids at some given temperature point. It even is known from Quantum Theory that as the temperature approaches absolute zero the Uncertainty Principle states that the gas could not occupy zero volume.

The idea of the zero volume concept was being developed by Gay-Lussac, but first mentioned by Lord Kelvin ( William Thomson ) in 1848. His work defined the temperature in terms of the $2^{nd}$ Law of Thermodynamics.

The following considers the Lifted Mass – Temperature relation of the Activity :

Since we are changing the temperature of the helium-filled balloon, and many of us have seen this, we expect some of the volume to change with temperature. Have you ever noticed a helium-filled balloon at different temps – the volume changes and its lifting ability of itself, let alone anything seems sluggish if not non-existent when cold.

If cold temps fail in this exercise you might consider warmer temps too. In this activity, one room or another in the house, or different times of the day may be warmer ( consider outdoor temps in the summer for example ). This may help the data table for this portion of the Activity.

All lifting ability is a direct measure of buoyancy as first noted by Archimedes ( see those two activities on the concept of buoyancy in a liquid – which applies to the atmosphere too ).

If a given helium-filled balloon can have an expected lift, why then can temperature affect this? With changing temperatures, the helium molecules will occupy either less volume ( in colder temps ) or greater volume ( at warmer temps ) which affects its overall density. If its density is higher in relation to air density, its lifting capacity which is directly derived from the difference in these two values then its lifting capacity will decrease, and if its density is less than its normal value, then inversely the lifting capacity will increase.

Our Activity looks for a relation but does not list one in particular. We expect some sort of relation, and we are searching for that outcome. Perhaps with research and investigation you find other relations.

Our activity uses simple things like a helium-filled balloon, a thermometer, needle and thread, and a measuring tape  along with a mass scale to investigate these ideas.

**Purpose :** To measure changes in volume of helium with changes in
temperature and compute the rate of change of volume
with temperature.

**Purpose :** To measure the lifting capability of a helium-filled balloon at given
temperatures to see if there is a relation of temperature to lifted mass.

## Materials :

- Helium-filled balloon(s) ( as round as possible – maybe 2 for validation or errors
  in experiment – also consider regular versus silver ones ),
- Marker,
- timer or clock,
- low-mass scale ( 0.1 g possibly even 0.001 g if available ),
- thermometers ( can be up to 3 lab quality -10°C to 110°C ),
- flexible measuring tape,
- sewing thread,
- sewing needle,
- marshmallows,
- refrigerator & freezer (or cooler with ice) ( each with space for he-balloon ),
- regular room at room temp,
- Slide Rule

## Procedure :

1) Be sure to follow all safety procedures. Have permission and supervision from an
   adult. For example once the needle is threaded, it is best to tie it off since it will be
   used in the whole of the Activity ( but an adult may be needed for this ). Do not
   mishandle hot or cold water. For all measurements handle the balloon efficiently (
   i.e. quickly ) so as to measure circumference when needed in the trials. As well to
   move through mass measurements of weight that the balloon can carry aloft at a
   given temperature of the balloon.
2) The directions given in the first few steps are the abbreviated version of the lab with
   only 3 data points for the graph. The steps after, note where other steps can be
   inserted if longer periods of time are available.
3) For the Helium-filled balloon, initially let it float in the room to be considered. Have a
   thermometer out to measure room temperature. Mark with a permanent marker a
   dot in 3 places around the balloon after wrapping firm but not squeezing with the
   flexible measuring tape. These points are used for each measurement for
   consistency.
4) After about 12 minutes ( this will be treated as the time frame for each location of
   the balloon ), record the room temperature and measure the circumference of the
   balloon. Record these values. ( T, C )

5) Attach the piece of thread to the balloon ( removing first any ribbon strand already attached ) with the threaded and secured ( ie tied off ) needle so that it hangs down.

6) Carefully attach marshmallow masses. The size of these will be determined by how much the balloon can just carry aloft. After each one is attached in turn let it go to see if the balloon can carry the marshmallow aloft. Once it cannot, subtract some as needed, just to the tipping point of where the mass of marshmallows lightly bounces on the ground.

7) Place the entire mass of string, needle and marshmallows on your zeroed out scale by holding the balloon so as to measure the entire mass. Record this value. ( M )

8) For the next two data points, we essentially want the helium-filled balloon to be at different temperatures. This can be accomplished in a number of ways and is up to you and the availability of resources.

9) For example, you have just used room temperature to measure your first set of points, why not use the outdoors ( if it is different ) – of course you will have to be cautionary with the balloon and have it secured most of the time so as not to float away ( letting it slip through your fingers as you attach masses for example ).

10) Another useful resource for different temperatures is a refrigerator / freezer – but again this depends on available space. Another cold spot might be a large cooler with ice in it. In each of these cases ( along with the outdoors ) the steps are essentially the same as #1-7 above.

11) Recognize that you have to take the helium-filled balloon out of the cold environment for measurements in most if not all cases, so time is of the essence – have a plan and work the plan to efficiently measure the measurements needed. Repeated exercises of this ( say three times per each situation ) and averaging these for a final value might be a good idea.

12) Note that the cold temps may result in failures in the lifting portion of the Activity – so what to do? If cold temps fail in this exercise you might consider warmer temps too. In this activity, one room or another in the house, or different times of the day may be warmer ( consider outdoor temps in the summer for example ). This may help the data table for this portion of the Activity. Garages, basements, and a sealed car may have different temps to consider. Note that for a car you need to have help and permission and do not remain in there with the balloon and thermometer. Always have permission and ask your parents for suggestions and help.

13) Calculations

14) With all the data measured at this point ( 3 sets ), the Calculations can commence :

15) At this point, the calculations can be done for radius, volume. Note, assume that the balloon is a perfect sphere.

16) Create a plot of Volume ( Y-axis ) versus Temperature ( X-axis ) and calculate the slope for the best fit line.

17) For a challenge compute the line of the graph and find the Temperature-axis ( x-axis ) intercept. See how close it is to -273° C, if using Celsius ( note : it is best to use Kelvin, so convert all temperatures to this ).

18) You can also use Charles' Law for each set of data ( V & T ) to see if their ratio is a constant value and how close they are to each other.

19) Determine the weight of each of the masses from the weight formula.

20) Now create graph of Mass versus Temperature. Is it a straight line or not?

21) Draw a best fit line and determine slope.
22) If the line appears rather curved, you might be able to analyze it through using a graph of the log(mass ) vs. the log(Temperature ), drawing a best fit line through this and determine slope. The values will give you a possible relation of the powers of these variables. ( Note : On the slide rule, this employs the L scale )
23) Note :
24) Though the Slide Rule is a recommended tool, all of the calculations can be done with a graphing scientific calculator or the use of a spreadsheet program. In these calculations you have to generate a table of data, graph it, and then find the slope and/or equation of the best fit line for the data. Other formula calculations can be done with these tools as well.

### Data :

| Trial | Location | Circumference (cm ) | Radius ( cm ) | Calculated Volume (cm$^3$) | Temperature (°C) |
|-------|----------|---------------------|---------------|----------------------------|------------------|
| 1 | | | | | |
| 2 | | | | | |
| 3 | | | | | |
| 4 | | | | | |

| Trial | Location | Circumference (cm ) | Mass lifted ( kg ) | Weight Lifted ( N ) | Temperature (°C) |
|-------|----------|---------------------|--------------------|---------------------|------------------|
| 1 | | | | | |
| 2 | | | | | |
| 3 | | | | | |
| 4 | | | | | |

### Calculations :

Use a Slide Rule for any and all calculations !

### Formula for determining radius 'r' from circumference 'C' :

$$C = 2*\pi*r$$

### Volume of balloon V :

$$V = \frac{4}{3}*\pi*r^3$$

$$m = \frac{\Delta Y}{\Delta X} \text{ ( useful for slope and finding intercepts ! )}$$

$$\frac{V_1}{T_1} = \frac{V_2}{T_2} \text{ ( Charles' Law )}$$

$$K = °C + 273$$

## For Fahrenheit to Celsius conversions :

$$°C = \frac{(°F-32)}{1.8}$$

## Weight :

$$F = m*g$$

( g = 9.8 m/s/s or 980 cm/s/s )

## Conclusion :

The first conclusion to consider is the conceptual behavior as expressed by the outcome in your numbers. How does volume change with temperature?

Depending on the calculations done, there are several ways to examine the outcome here. If you used Charles' Law, you can compare ratios to see how similar they are. With a graph, the Temperature-intercept can be found and examine how close it is to the expected value of -273°C.

The next set of conclusions comes from lifting capacity of the balloon. Think for a moment of hot-air balloons in operation. Do they merely heat the air once or as they sail along from time to time and what does this do for the balloon and passengers? Did you find the same thing?

Note one can estimate the buoyant force for the lifting capability of the balloon from the following expression ( note to look up the density of both air at STP and helium ) :

$$F_B = g*V*( D_{Air} - D_{Helium} )$$

# Activity #22
## Cartesian Diver Fun Density Determination Activity
Grade Level : High School
Math Level : Calculating

In this Activity we construct a very simple item called a Cartesian Diver. One of the critical points involving the science of the Cartesian Diver is Density, so it is here we will concentrate on its Density and how it varies in the operation of the Cartesian Diver's operation.

Density is a very important and early-on understood concept in human history for the physical properties of materials. Anyone who has picked up a stone and a sponge of roughly the same volume can testify to as to which is the more massive and therefore the more dense.

Density by definition is Mass per Unit Volume :

$$\textbf{Density} = \frac{\textbf{mass}}{\textbf{Volume}}$$

$$\rho = \frac{m}{V}$$

Density is one of the characteristics of a material to help in its determination ( when there are few other means to consider ). Metals have large values of density, such as aluminum at 2.70 g/cc and iron at 7.80 g/cc. Other than metals, other objects can have measured density values, such as water ( 1.0 g/cc ), typical glass ( 2.60 g/cc ), common wood ( 0.70 g/cc ), ice ( 0.92 g/cc ).

The majority of information about Density is found in the Density Activities on www.cosmicquestthinker.com . More about it in the reading and in terms of Activities is located there. Also, Archimedes' Principle has density involved in it.

In this Activity we construct a Cartesian Diver. With careful observation and measurement, we can determine its density without water in it, with enough water in it to allow it to float, and with enough water to make it sink.

A Cartesian Diver is a toy or experiment is where there is a small rigid tube open at one end, much like an eyedropper which when it has enough water and enough air in it and placed in a flexible container ( such as a pop bottle ) with water, so that it floats. It is positively buoyant. That is to say its present density is less than that of the surrounding medium ( the water ).

When the container is closed ( the lid is put on the pop bottle ) and a force is applied to the container ( such as squeezing the pop bottle ) transmits the force through the fluid ( water ) so that some more of it is forced into the rigid tube ( the eyedropper ) so as to increase its mass, hence its density. With a density greater than that of the surrounding medium ( the water ) it can now sink. Removing the force allows the water to leave the tube ( the eyedropper ), decreasing its mass, hence its density, and it returns to floating.

## Pictures of Materials and the Cartesian Diver :

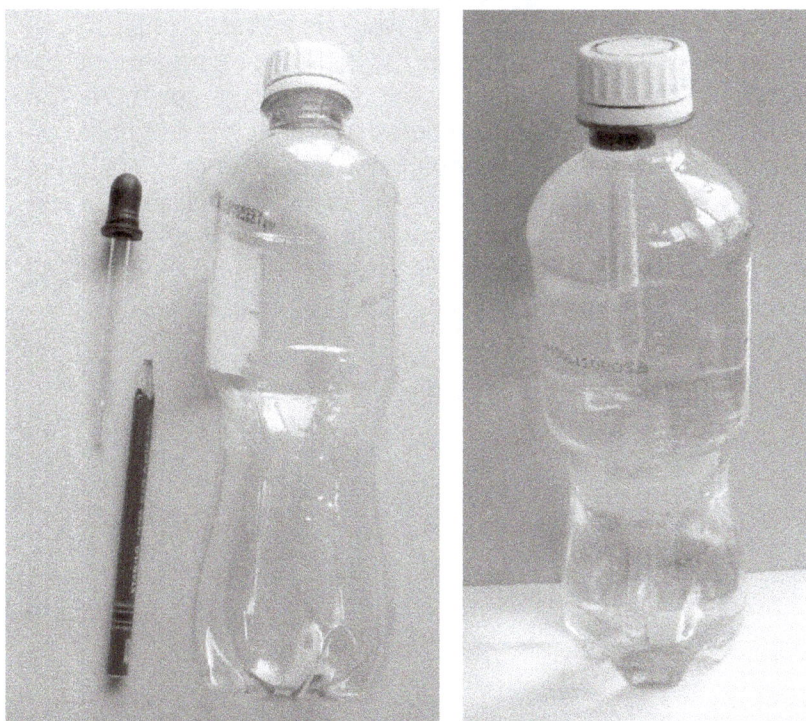

Despite its simplicity, there are a host of scientific rules and laws that apply to the **Cartesian Diver**. To begin with, the name itself for the device comes from **Rene Descartes**, the renowned mathematician who used a similar device made from materials appropriate to the technology of his day to demonstrate the principle of **Buoyancy ( as noted in Archimedes' Principle )** and the **Ideal Gas Law**. Clearly Buoyancy is involved. The diver goes from being positively buoyant to negatively buoyant since its density goes from less than water's density to greater than water's density. To achieve this, a Force is applied, hence invoking **Newton's Laws** as well. Further the applied force operates as described by **Pascal's Law**. Pascal's Law states : The applied force to the fluid ( water ) is transmitted undiminished throughout it and the pressure ratio remains the same. Why things change are due to the nature of the natural behavior of **States of Matter**. In the case of the **Liquid**, it takes

223

the shape of its container and cannot be compressed, whereas the **Gas**, air in this case trapped in the eyedropper, which takes the shape and volume of its container is readily compressed. The force on the water transmits through it and since the air is compressible the liquid compresses the air, increasing the mass of water in the eyedropper, hence increasing its density. Obviously, the volume of the gas is changing in this experiment and this is an application of **Boyle's Law**, which states that the absolute pressure and volume of a given mass of confined gas are inversely proportional ( if temperature remains unchanged within a closed system – which we effectively have here ). This essentially means that the product of Pressure and Volume equals a constant. In the case of the Cartesian Diver, we apply a Force, which is a Pressure. Since Pressure increases, the Volume must decrease by the inverse-proportion to the pressure increase to remain constant, so then, the volume of the gas decreases. This product, pressure times volume is the beginning of the **Ideal Gas Law** equation. This equation is a good approximation of the behavior of many gases under many conditions and relates the key ideas of Pressure, Temperature, Volume, and the Amount of a Gas : P*V = n*R*T ( P = Pressure, V = Volume, T = Temperature in Kelvins, n = the number of moles of a substance ( amount of substance ), R = the Gas Constant ).

**Purpose :** To use a measured masses from an eyedropper with no water, with enough water to allow it to float, and enough water to allow it to sink in a Cartesian Diver and the determination of its Volume from volumetric-displacement in a graduated cylinder to calculate the Density at each of these stages.

**Purpose :** To use a measured volume and measure mass from a given quantity of water, graph the mass vs. volume and from slope determine the density of water.

Note : The density of water can be either assumed to be known, or done with the directions here or as taken from the Density Activities page on the web site. The density of water is the follow-up directions in the If Needed section of the Procedure.

## Materials :

- Mass Scale ( sensitive to 0.01 g if possible, minimally 0.1 g ),
- Graduated Cylinder ( minimum 25 mL size – be sure the eyedropper used can fit into it easily ),
- Glass Eye Dropper,
- Plastic Pop Bottle ( 16 oz. or up to 2 L size ),
- Wax Marking Pencil,
- Regular Cup,
- Water,
- Slide Rule

## Set Up & Construction Procedure of the Cartesian Diver :
## The Changing Density of the Cartesian Diver :

1. Set Up & Procedure :
2. First measure the mass [ m ] of the Eyedropper with no water and record this value in the Data Table.
3. After acquiring materials, clean the pop bottle thoroughly and let it dry.
4. Fill the Pop Bottle with water to the area where the cap screws on.
5. Nearly fill the glass being used with water as well.
6. Test the Eyedropper and make sure it works properly.
7. Fill the Eyedropper with as much water as you can. It should now sink in a cup of water.
8. With the Eyedropper full of water, its Volume [ V ] can be determined :

9. First place enough water in a Graduated Cylinder so that the Eyedropper would be covered ( Note it is important that the Eyedropper can go in and out of the Graduated Cylinder, so be sure to have the right size one ). This amount of water is the Initial Volume [ $V_i$ ]. Record this value in the Data Table.

10. Place the water-filled Eyedropper in the Graduated Cylinder with water and see to where the new level is. This is your Final Volume [ $V_f$ ]. Record this value in the Data Table.

11. To see for yourself, place the Eyedropper in the cup of water to see that it awkwardly floats. You need to put some ballast ( weight ) in it – that is, the water.

12. Let the Eyedropper take in a small amount of water and now place it in the cup of water. It should float reasonably well.

13. Adjust the amount of water in the Eyedropper so that just the top of the Eyedropper top is just at the top of the water in the cup.

14. Carefully lift it out and mark the water level in the Eyedropper with the wax marking pencil.

15. Now carefully measure the total mass of the Eyedropper with this amount of water that allows for neutral buoyancy.

16. Note that it can be difficult to have an Eyedropper on a Mass Scale with water in it, so be patient and practical. Zeroing out a small item so as to lift the end where the water can come out and the Eyedropper is now level is best.

17. It is best to do this neutral buoyancy step 3 times and then average the results and place the average amount for mass [ m ] in the Data Table.

18. Construction of the Cartesian Diver :

19. Now with the Eyedropper with the amount of water needed to achieve neutral buoyancy place it in the pop bottle which has water in it and close it off by putting the cap firmly on.

20. Test the Cartesian Diver by squeezing the bottle ( applying a Force ). If done correctly, the Eyedropper will fill with water and the Cartesian Diver will sink. When you release the force, the liquid returns to its original place and levels, so the Cartesian Diver should then rise back to the surface of the water.

21. Some of the observations to make are these in the form of leading questions : How does the Cartesian Diver descend and ascend? That is, is it similar to when one drops an object in air ( in a straight line )? ( Tilt the bottle slightly when testing your hypotheses ). Does it ascend or descend at a rate like the acceleration due to gravity – faster or slower? How could you determine this?

22. For our Activity, now carefully observe the Cartesian Diver as you slowly apply the force needed to fill it with water. You want to observe the place that the water rises to when it sinks.

23. Take the Cartesian Diver out and mark this spot where the water needs to go to.

24. Place the Cartesian Diver back in the bottle and test this new mark on the Eyedropper to see how accurate you are. Correct this as needed until you just match the place.

25. Now remove the Cartesian Diver.

26. Fill the Eyedropper with enough water to match the mark needed to allow the Cartesian Diver to sink.

27. Measure the total mass of the Eyedropper with this additional mass of water.
28. Take 3 measurements of the Eyedropper with this additional mass of water and average the results. Record this average mass [ m ] in the Data Table.
29. Using the various Mass Readings ( you should have 3 – A) empty, B) enough water for neutral buoyancy, and C) maximum water to sink ) in combination with the Volume to determine the Density [ ρ ] of the Cartesian Diver. Should the Density increase or decrease? How should it compare to the Density of Water?

## If Needed Activity : Density of Water :

1) Place an empty graduated cylinder on the mass scale and zero it out.
2) Pour into the graduated cylinder a measured amount of water ( for example 3.0 or 5.0 mL )
3) Record the Volume of Water ( V ), and measure and record the mass of the water in the graduated cylinder ( m ).
4) Add more water to a new level ( for example, if you started with 5.0 mL, increase to 8.0 mL )
5) Record the total mass ( m ).
6) Measure and record the new total mass of the system with the new volume measure ( V ).
7) Continue to add volume and measure mass so as to do at least 4 trials.
8) Graph the date with mass ( m ) on the y-axis and volume ( V ) on the x-axis.
9) Draw the best fit line and determine slope of the line, which is the average density ( ρ ) for water.
10) Compare this value to the generally accepted value for water ( 1.0 g/cc ) and determine the experimental error

## Data :

## Activity : Changing Density of a Cartesian Diver

Mass of the Eyedropper  [ m ] :

| | |
|---|---|
| Eyedropper with no water Mass ( g ) | |
| Eyedropper with enough water to have neutral buoyancy Mass ( g ) | |
| Eyedropper with maximum water needed to cause it to sink Mass ( g ) | |

Volume of the Eyedropper [ V ] :

| | |
|---|---|
| Final Graduated Cylinder reading  [ $V_f$ ] | |
| Initial Graduated Cylinder reading  [ $V_i$ ] | |
| Change in Volume [ ΔV ] – the Volume of the Cartesian Diver | |

<u>If Needed Activity : Density of Water</u>

| Trial | Total Volume ( V ) | Total mass ( g ) |
|-------|--------------------|------------------|
| 1     |                    |                  |
| 2     |                    |                  |
| 3     |                    |                  |
| 4     |                    |                  |

## **Calculations :**

Be sure to use your Slide Rule

### **Calculate Volume :**

Change of Volume ( Volume of Object ) =
Final Volume in Grad. Cyl. – Initial Volume in Grad. Cyl.

$$\Delta V = V_f - V_i$$

### **Average Mass Value :**

$$\textbf{Average Mass Value} = \frac{\textbf{Sum of Total Mass of all Trials}}{\textbf{Number of Trials}}$$

### **Calculate Density :**

### **Density Formula :**

$$\rho = \frac{M}{V}$$

### **Slope :**

$$m = \frac{\Delta Y}{\Delta X}$$

### **Graphing Mass ( x-axis ) vs. Volume ( y-axis ) Slope :**

$$\textbf{Density (Slope)} = \frac{\textbf{change of mass}}{\textbf{change of Volume}}$$

<u>If determining the Density of Water :</u>

Accepted Value : Density of Water = $1.0 \frac{g}{cc}$

$$\%Error = \frac{[\,accepted\,value\,-\,activity\,value\,]}{accepted\,value} * 100\%$$

## Conclusion :

What was the Density of the Cartesian Diver with no water? What about The Density of the Cartesian Diver with enough water in it to make it neutrally buoyant? What about the Density of the Cartesian Diver with the maximum water amount so that it just sinks? How do these values compare to the accepted value of Water?

# Activity # 23
## Complete Hooke's Law Activity
Grade Level : High School
Math Level : Calculating


At one time or another we have all bent a flexible material, such as plastic or metal, and let it go. As we pulled it, we notice there is a force pulling back on us. We find that it moves opposite to our original motion, from a restoring force. Typically the object will move quickly right past its original starting position and move almost as far in the opposite direction that it was originally moved. Once there it slows to a stop and moves back in the opposite direction again. The mass is vibrating back and forth past its starting point, called the equilibrium position. Given time it once again comes to rest at this point.

The motion of the object has zero speeds at its extremes and maximum speed at the equilibrium point in the center. Since it is changing speed, it is therefore accelerating. This means a force is acting on the mass. It is often referred to as the restoring force.

Also any change of speed means a change in the kinetic energy of the object. This means that work is being done or by that system. First work had to be done to move the mass ( can be a stretched or compressed spring, a stretched rubber band, or plucking a taught wire ). This initial work is a change in elastic potential energy of the system. This in turn becomes the change in kinetic energy of the system. The reason it does not move indefinitely means that there must be a force acting on the system ( according to Newton's Laws of Motion ). This force is friction, which brings the system to a stop and its energy, which cannot be destroyed is turned into heat and sound. The decrease in motion is called damping.

A key question arises, however. What does this restoring force depend on? The answer came from Robert Hooke in 1678 when he described the restoring force and its dependency on the displacement of the mass. That is to say, the more one stretches a spring, the greater the restoring speed. The force, however, is only proportional to the displacement. To make it an equality, a constant of proportionality is needed, here called the spring constant. The value of this is not universal for any and all springs, but instead depends on the spring. The harder it is to move the spring, the stiffer it is, hence the 'k' value is greater. It's units are N/m, which can be seen from the formula described here :

Hooke's Law

$$F = -k * \Delta x$$

The spring constant is a characteristic or parameter of an elastic materials, such as a spring ( for which it was originally named ) to represent a measure of a spring's resistance to being stretched or compressed.

Note, it is important to realize that too much stretching can extend an object beyond its elastic limit, and will render the spring basically permanently sprung and in the case of a rubber band, typically broken.

Also in the case of the Activity, springs within a given range of stretching will be very consistent, while rubber bands have as many as 3 different ranges over which they have 'k' values, so in obtaining data be careful with rubber bands and keep the ranges small for a reliable figure for the calculation of its 'k' value.

The application of this motion can be seen by considering simple harmonic motion. Simply defined, simple harmonic motion is where there is a vibration about an equilibrium position in which a restoring force is proportional to the displacement from equilibrium. This type of motion needs to be understood and controlled in all spring-shock systems for vehicles or anything that needs to remain stable when in randomly-oscillating, or bumpy situations. It also is parallel to Pendulums since both exhibit the same characteristic back and forth motion where the force is a max at the extremes and is zero at equilibrium. One key difference comes from the formulas for determining the period of a pendulum as compared to a mass-spring system. The simple pendulum does not depend on mass as the amount of restoring force ( gravitational force ) here increases as the mass, hence inertia, of the pendulum does. The mass-spring system on a level surface and only relying on the force due to the spring, has a situation where the mass's magnitude does not affect the restoring force. An increase of mass here results in a smaller acceleration, hence a greater time to complete one oscillatory cycle. Never-the-less, the formula for the Period of a Spring-Mass system is comparable to a regular Pendulum as can be seen in this formula :

Period Formula :

$$T = 2 * \pi * ( m/k )^{1/2}$$

Other types of mathematical applications for restoring forces look at bow strings for archery, pendulums in clocks or when acrobats such as trapeze artists are in motion.

Elastic materials can also store energy. In one part of this Activity we examine the energy of the rubber band when stretched. Here it has what we call Elastic Potential Energy. Anyone who has shot a rubber band or stretched and released a spring can readily note the motion of said object which came from imparting energy into the elastic material where it was temporarily stored until released and turned into kinetic energy of motion as well as heat and sound.

Potential Energy can come in a variety of forms and is all due to the properties or conditions of the matter. Elastic Potential Energy can be stored in any solid material that can be deformed and it then returns to its original state when released, such as a stretched or compressed spring, a stretched rubber band, a bounced super ball, or any other rubber or elastic ball for that matter. It does not always have to be things that are made of rubber or plastic, however.

Virtually all materials have a certain amount of response or return to its natural 'at-rest' state when distorted. Take, for example the fact that most items bounce some when they hit the ground after being dropped. This is due to their flexing and returning to their original form.

Stored elastic potential energy has a number of uses. Take into consideration the use of springs or bungee cords. The simplest, most common we encounter are spring-activated toys that pop back after being compressed. There are rubber elastic counterparts to this, such as the poppers, which are essentially half of a hollow tennis ball that one inverts and places on the ground and then it pops back into shape and in snapping against the ground launches into the air.

Like all energy forms, elastic potential energy can only be stored or converted into other forms of energy. Like all energy it cannot be created nor destroyed.

One of the Activities on the web site www.cosmicquestthinker.com is Elastic Potential Energy where it too utilizes Hooke's Law and looks to rubber bands to measure this type of energy. As with any energy form, elastic potential energy can be converted into other forms of energy ( this is the purpose of the Elastic Potential Energy Activity ). In this present Activity we only use the known information as determined in the course of the Activity to calculate the expected Elastic Potential Energy value.

Formula for Elastic Potential Energy :

$$PE = \frac{1}{2} * k * x^2$$

The activity here has several steps to the Purpose. In the first, it examines the determination of the spring constant for a variety of springs and for rubber bands as well. In the follow-up Activity, there is the possibility of using this information for further calculations involving the elastic potential of a spring-mass system, hence an ability to estimate through calculation a predicted speed a mass is moving at maximum speed in its oscillation. Also in the third aspect of the Activity, the Period of an oscillating spring ( or rubber band ) with a mass on it will be measured and compared to the calculated expectation as well.

**Purpose :** To determine the spring constant, k, for a spring
( or elastic material ( i.e. rubber band )) through
increasing force and measured displacement values.

**Purpose :** To determine the Elastic Potential Energy, PE, for a given spring
stretched a given distance.

**Purpose :** To determine the expected Period, T, for an oscillating spring with a
mass attached when stretched then released and allowed to oscillate.
This value is compared to the results of measurements of the
Period, T.

**Purpose :** To calculate the expected maximum speed, $v_{max}$, of an oscillating
spring-mass system when it is oscillating.

## Materials :

- Springs ( small ) or
- Rubber Bands ( buy bags of specific types for uniformity, 3 varieties is good if consistency is desired ),
- Best Everyday Mass Ideas : can use a eyebolt with nuts that can attach ( measure mass of each part separately ),
- Mass Scale,
- Meter Stick and/or Ruler,
- Tape,
- Stack of Books or Chairs,
- Dowel Rod,
- Stopwatch,
- Twist Ties,
- Slide Rule

Note : If no eyebolt, then use an ordinary long bolt with nuts. Attach the bolt
with twist ties securely to act as the eyebolt loop.

Note : In the following picture is a basic bolt altered to be an eyebolt and using
twist ties to make the loop and the pointer. Note that one can use the
eyebolt or rubber bands instead. Also it is up to you in deciding to use
chairs or stacks of books to allow for the set up of the system. In the set
up, chairs and books will simply be referred to as the vertical system.

**Picture of Most Basic Materials :**

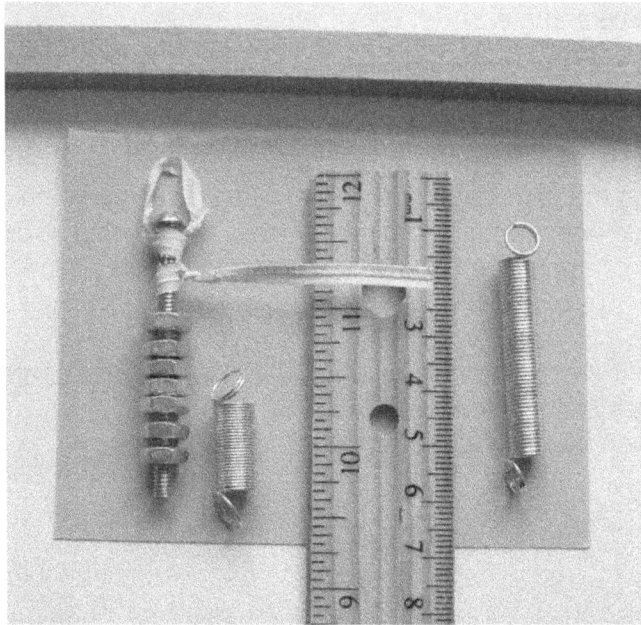

**Set Up & Procedure : For regular Springs and Rubber Bands**

1) Set Up :
2) Note : All measurements need to be to the nearest 0.1 g and 0.1 cm.
3) Know the masses being used by measuring them on the mass scale. Record these results in a table individually – it is best to create a labeling system for them, such as 1, 2, et al or A, B, etc.
4) The best mass system is using an eye hook bolt with nuts. Measure each separately.
5) A convenient stand for the rubber bands and/or springs is having 2 stacks of books near each other and bridged by a dowel rod. Make it level and tall enough so that when the rubber bands or spring attached to the dowel rod by twist ties and hanging vertically is then elongated by the addition of masses allows for it to stretch. Another good option is two chairs and having the dowel rod lying across their backs where the dowel rod taped into place. When referenced in future steps both the books and chairs are called the vertical system.
6) The dowel rod can be wood and can be notched slightly to let the spring ( or rubber band ) hang down from the rod. Note : It does not have to have a notch and this should be done by a parent.

7) Arrange a ruler/meter stick nearby so as to stand vertically and close enough so that a twist tie coming horizontally from the mass attached to the spring ( or rubber band ) would be readable along the meter stick/ruler. – In the case of the eyehook the horizontally attached twist tie at a chosen point and it remains constant so that even when the eyehook is off it is pointing at the ruler/meter stick. At the bottom of the spring ( rubber band ) is the best, since additional masses will be added as the Activity progresses. – Since the pointer is short it is best to have the vertical system ( i.e. books or chairs ) near each other. This also allows for a measuring tool to be near to stand.

8) First record the initial position of the bottom of the spring ( or rubber band ) according to the scale, with no masses ( eyebolt included ), attached. Note this does not have to be 0, but it is best if arranged so this is the case. However make note of the fact that at one point is the Activity the mass on the spring will oscillate up and down and we need both the highest value and the lowest value, so it may be better to start at some other number, say 2.0 cm for example as the best value. In that case displacement is found by the absolute value of the difference of the two values on the ruler.

9) The Activity :

10) Now Attach the eyebolt hook to the system ( spring, rubber band ). Be sure that it does not touch the scale and is parallel to the table top.

11) Record both the mass of the eyebolt and the new position of the spring ( or rubber band ). Be sure to always measure to the same initial spot of measurement ( i.e. the bottom of the spring or rubber band )

12) In each trial then add the additional masses ( nuts ( all known masses ) ) sequentially.

13) Note : With each trial record the total mass [ $m_T$ ] for the system. This is not just the added mass, but the total on the system at that trial.

14) For the given trial measure the total displacement [ $\Delta x$ ]. Note this is not the incremental displacement, but the total displacement.

15) The data table should read with increasing masses and increasing displacements if done correctly.

16) Go through all of the masses. It is best to have 4 or 5 masses to add.

17) If doing more than one spring ( rubber band ), then perform each of these steps of adding masses and recording the results again.

18) Calculations ( for each spring separately ) :

19) Calculate the weight of each of the masses used for each trial [ $F_i$ ]

20) Convert the Displacement from cm to m, then

21) Graph the points of Force [ F ] on the y-axis vs Displacement [ $\Delta x$ ] on the x-axis.

22) Draw a best fit line, which should be linear. Determine the slope of the line which is the Spring Constant [ k ] for that spring ( rubber band ).

23) Follow all of these steps for each of the springs, rubber bands used.

24) Further Calculations to be done :

25) Any stretched spring has Elastic Potential Energy [ PE ], which we can determine :

26) Using the Potential Energy formula provided, calculate the Elastic Potential Energy for each of the Displacements that resulted from the added masses in the trials. Make a separate data table for this.

27) Look to the Conclusion for other things to do with the Elastic Potential Energy values ( hint you may need the L scale of the slide rule ).
28) Continuation of Activity & Calculations :
29) For at least 3 masses on the spring system go back and redo the above Activity for measuring the total mass and displacement.
30) With each of these trials separately, now stretch the spring slightly and let it go.
31) As you let the spring-mass system go, start a stop watch and measure the amount of time for a chosen number of complete oscillations [ N ]. A complete oscillation is when it goes all the way to its highest point and returns to the starting position you had stretched to at the bottom [ S ].
32) It is easiest to choose a number easy to mentally divide by, such as 10, but for those who want to employ the slide rule, choose another number, such as 5 or 8, or something else.
33) Also watch as the system oscillates and note the top and bottom numbers on the ruler/meter stick and write these down. The difference [ R ] of these numbers ( we will call A & B ) is twice the maximum displacement of the system [ $x_{max}$ ] which we then determine. Note that estimating these values to the nearest $1/10^{th}$ of a cm will be difficult at best, so whole numbers or half cm values may be the most realistic possibilities.
34) Note for any given trial, it should be done at least twice if not three times and the results for that given trial are then averaged. Use the Average Results of A & B, as well as S in your final calculations.
35) From your values of N and average S, determine the Observed Period [ $T_O$ ].
36) Calculate the Predicted Period [ T ] for each of the trials you have conducted.
37) Compare your Observed Period, $T_O$, to the Predicted Period, T. How similar are they? Why are they different?
38) Determine average values for A & B. Use these to determine the maximum displacement [ R ] and then use this to find $x_{max}$.
39) Use the maximum displacement, $x_{max}$, to calculate the predicted speed, $v_{max}$, of the oscillating spring-mass system for each of these trials.

**Data :**

| Mass Label | Mass [ m ]( g ) |
|---|---|
|  |  |
|  |  |
|  |  |
|  |  |
|  |  |

| Trial | Total Mass [ $m_T$ ]( g ) | Weight [ $F_T$ ] ( N ) Calculated | Displacement place [ $\Delta x$ ] (cm ) | Displacement [ $\Delta x$ ] ( m ) Converted |
|---|---|---|---|---|
| 1 |  |  |  |  |
| 2 |  |  |  |  |
| 3 |  |  |  |  |

Chosen Number of Oscillations [ N ] : _____

Amount of time [ S ] for N number of oscillations : _____ s

Calculated Period of Spring-Mass System from Observations : _____ s

Predicted Period of Spring-Mass System from formula : _____ s

## Calculations :
Be sure to use your Slide Rule!

### Calculation Steps :
Note : Mass has been taken in grams in this exercise.

Convert all mass units from grams into kilograms ( 1 kg = 1000 g )
Convert all distance units from centimeters to meters ( 1m = 100cm )

### Hooke's Law

$$F = -k*\Delta x$$

Rearranged as ( taken as absolute value ) :

$$k = \frac{F}{\Delta x}$$

## Force ( Weight ) Formula :

$$F = m*g$$

$$g = 9.8 \text{ m/s}^2$$

## Predicted Period of a Spring-Mass System :

$$T = 2*\pi*( m/k )^{1/2}$$

## Observed Period Formula :

$$T_O = \frac{S}{N}$$

## Predicted Max Speed of the Oscillating Spring-Mass System :

$$v_{max} = \Delta x_{max} * ( m/k )^{1/2}$$

Note : $x_{max}$ is found from this formula set :     $R = A - B$

( Total Displacement in Oscillation is the absolute value of the
difference of the two points on the ruler/meterstick  - also A & B
are the average values from at least 2 if not three separate times
using the same materials )

$$x_{max} = \frac{R}{2}$$

## Elastic Potential Energy :

$$PE = 0.5*k*(\Delta x)^2$$

## Average :

$$X_{ave} = \frac{\Sigma x_i}{n}$$

## Slope :

$$m = \frac{\Delta y}{\Delta x}$$

## Conclusion :

If you have done more than one spring / rubber band how do the spring constants compare? For example, the easier it is to stretch a spring, is the 'k' value smaller or larger than a spring with a larger 'k' value?

Looking at the Elastic Potential Energy values you have calculated ( and looking at the formula ) as the displacement changes, how does the energy value change? Does it change linearly or in some other fashion ( quadratic for example )? How do you know? Note — you can find this answer by graphing the log of the Potential Energy value and graphing it against the log of the displacement and drawing a best-fit line and finding its slope. The slope will be the power relation of the energy to the changes in displacement.
How do your Observed and Predicted Period values compare?
Do your maximum speed values seem reasonable or not?

# Activity #24
## Elastic Potential Energy of a 'Popper' Toy Activity
Grade Level : High School
Math Level : Calculating


This Activity explores a toy often called a Popper, which looks like a tennis ball cut in half. You invert it, putting elastic tension on the rubber body and can either place it on the floor or table top or drop it where it releases its elastic potential energy turning it into kinetic energy which moves it into motion that turns the kinetic energy of motion into gravitational potential energy. At some point it reaches its maximum height ( all the elastic potential energy has gone into kinetic energy which has now turned into gravitational potential energy ) and then falls back to the Earth.

Elastic Potential Energy is energy due to the condition or tension in the molecules of a system when they are moved from their rest or minimal energy position into a place of tension. This is much like a stretched rubber band, or a stretched or compressed spring. Even a wooden or plastic ruler held straight and one end is slightly pulled back has elastic potential energy, much like a bow string of a bow.

There is more information about energy in the Hooke's Law Activity – so read that prelude to increase the ideas of the Popper.

This Activity uses the Popper, lets it go, measures the height to which it rises, and uses the Gravitational Potential Energy to be representative of the Elastic Potential Energy of the item ( though , in reality, some is lost due to some of the molecular bonds breaking, sound, heat, and other energy dissipation ). From this, we can estimate the spring constant 'k' of the rubber material ( see Hooke's Law ) and the amount of force ( average ) exerted to distort and place in dynamic tension the popper. Also we can use Conservation of Energy to estimate the speed of impact ( hence speed of take-off ) of the popper at the ground through Kinetic Energy considerations.


**Purpose :** To approximately measure the Elastic Potential energy of a Popper Toy through examination of the Gravitational Potential Energy of the toy.

**Purpose :** To approximately estimate both the 'spring constant' and average force imparted to the Popper toy from the estimated energy values.

## Materials :

- Popper ( good to have a couple different sizes if possible ),
- Ruler,
- Meter Stick,
- Digital Camera with Video capability ( Only IF desired and not absolutely needed but can be helpful in determining the height of the popper's attained altitude ),
- Mass Scale ( sensitive to 0.1 g ),
- Goggles,
- Different surfaces to have popper operate from ( tile or hard floor, carpeted floor or with a cushion ),
- Slide Rule

## Pictures of Poppers :

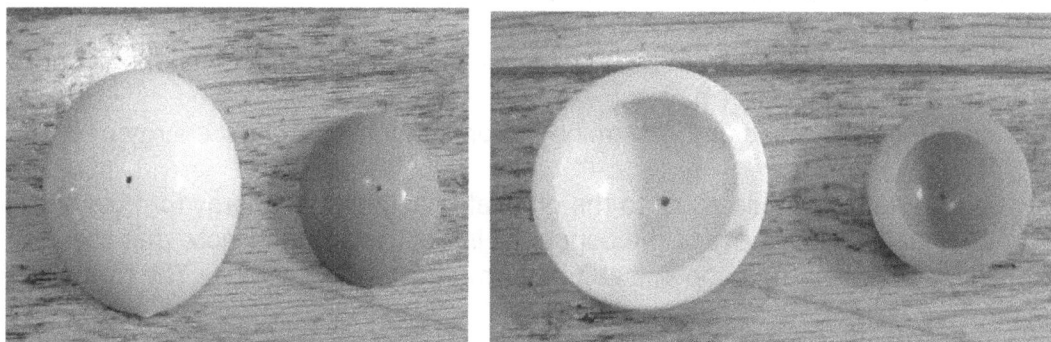

## Procedure :

1. Note : You should wear goggles and avoid being directly in front of the poppers.
2. If possible try to use 2-3 poppers ( time permitting ) where each is separately measured. Note, it is best to have at least 2 different size poppers ( see photo ) for comparative purposes ( not required, activity can be done with one popper ).
3. Measure the mass ( m ) of the popper on the scale to the nearest $1/10^{th}$ of a gram and record this value.
4. Using a ruler, measure the total distance the popper is distorted ( from the top of the dome when the relaxed position to the inversion of the popper when elastically tense ). This should be measured to the nearest 0.1 cm. Realize this too is an estimated value, so repeated efforts at this may be a good idea and average the results. Record this value ( x ).
5. First do 1-3 trial runs to obtain a visual of the popper's motion and maximum height it rises to. Set it on the floor in the tensed configuration so that when it pops, it propels into the air ( typically this means dome side away from the floor ). Doing this near a wall, place a meter stick vertically ( and held in place with a piece of tape or two ) to watch and see how far the popper goes vertically.

6. If less than one meter, then the system you have will work. If greater than one meter you need to either use a second meter stick taped above the first meter stick vertically or if only one meter stick, then at the place where the first meter stick measures to now move this meter stick and tape it in the place at 1 m from the floor.

7. Perform enough trials to obtain a visual estimate of the maximum height of the popper. It is best to look as directly at it as possible ( if too far up, this may affect your values – you could consider with parental permission and supervision the use of a small step device to add a few inches to your field of view )

8. An interesting alternative to determining the height to which the popper goes is to use a camera that has video capabilities. With some time, practice, and proper use you can have the camera focused on the place where the popper goes to you can examine frame by frame the rise and fall of the popper and determine its final height.

9. With the method in place and an idea of the height attained by the popper, do several trials ( at least 3 and as many as 10 – the more you have time for the better, since this will create a good set of data to generate an average value from ).

10. Do all the trials and record all the height values ( h ).

11. Calculations :

12. Determine the Average height from the heights measured in the trials – total of heights divided by the number of trials ( have ).

13. Use this average height value and determine the gravitational potential energy of the popper ( $PE_g$ )

14. Set the Kinetic Energy formula equal to the gravitational Potential Energy formula and solve for the final impact velocity of the popper as it hits the ground ( note that the initial speed is zero when the popper is at its maximum altitude ). ( Hint : The formula will simplify to $v = ( 2*g*h )^{1/2}$

15. Assuming that all of the elastic potential energy is translated into gravitational potential energy, now determine the 'spring constant' 'k' for the popper. ( Note that the distortion distance 'x' is used here ). ( Hint : Formula will simplify to : $k = \dfrac{2*PE}{x^2}$ ).

16. Using the determined 'k' value, calculate from Hooke's Law the average amount of force ( F ) needed to move or bend the popper the distortion distance ( x ).

17. If you have other poppers, follow the same directions for this as well, recording their results separately. Also compare the results for large and small poppers.

18. Another alternative is to use different surfaces ( cushioned ) for comparison.

## Data :

Measured distortion of Popper [ x ] : _____ m

Mass of Popper [ m ] : _____ kg

| Trial | Height of Popper [ h ] ( m ) |
|---|---|
| 1 | |
| 2 | |
| 3 | |
| | |
| Last trial | |
| Average [ $h_{ave}$ ] | |

## Calculations :

Be sure to use your Slide Rule!

Let Acceleration due to Gravity be g = 9.8 m/s$^2$

Average :

$$X_{ave} = \frac{\sum_{i=1}^{n} x_i}{n}$$

Gravitational Potential Energy :

$$PE_g = m*g*h_{ave}$$

Elastic Potential Energy :

$$PE_{el} = \frac{1}{2}*k*x^2$$

Hooke's Law :

$$F = k*x$$

Conservation of Energy :

$$\Delta PE_g = \Delta KE$$

$$\Delta E = E_f - E_i$$
The change of Energy is the final value minus the initial value

Kinetic Energy :

$$KE = \frac{1}{2} * m * v^2$$

## Conclusion :

Do you think all of the elastic potential energy went into the gravitational potential energy of the popper, why or why not? What other forms of energy could have the elastic potential energy been turned into?
How do large and small poppers compare for spring constant values, for maximum height values, for gravitational potential energy values?

# Activity #25
## Elastic Collision with Conservation of Momentum & Energy Activity
Grade Level : High School
Math Level : Calculting

The main ideas contained in this Activity are discussed in the Activities concerning Mechanical Energy ( which is an online Activity as well as being found in the first volume of this series ) and the Conservation of Momentum Activity ( Activity #32 ) in this book. Here the foundations of Science and Physics are noted, where the Conservation of Energy and Momentum are outlined.

In the case of an Elastic Collision ( where the impacting particles remain intact and are not conjoined ), both the Momentum and Energy are Conserved. This means that the initial amount before the impact is equal to the outcome value after the collision.

As noted in the Preludes, Energy, like Momentum, cannot be created nor destroyed. The amount that a system has going into an interaction is the maximum amount that can come out of it.

**Energy$_{Before}$ = Energy$_{After}$**

**Momentum$_{Before}$ = Momentum$_{After}$**

Energy is not conserved in Inelastic Collisions because some of the Kinetic Energy turns into heat, sound, and other forms of energy in the process. In both Inelastic and Elastic Collisions, however, Momentum is conserved.

To solve an Elastic Collision case, as it is in this Activity, we not only have to consider the Conservation of Momentum of the situation, but also the Conservation of Energy. Hence we have not one, but two equations to solve.

$$0.5*m_1*V_{1i}^2 + 0.5*m_2*V_{2i}^2 = 0.5*m_1*V_{1f}^2 + 0.5*m_2*V_{2f}^2$$

$$m_1*v_{1I} + m_2*v_{2I} = m_1*v_{1I} + m_2*v_{2i}$$

To simplify the situation here, we will consider a one-dimensional case and have the initial source of energy and momentum be only one of the objects ( the impactor ball ) while the other remains at rest initially ( the impacted ball ). Also we will have each of the masses as being as close as possible since differences in mass will result in still more variables to deal with ( that is to say, the Impacting Ball will have both kinetic energy and momentum after the collision. Ideally if the two masses are the same, it should transfer all of its energy to the other ball ).

$$0.5*m_1*V_{1i}^2 = 0.5*m_2*V_{2f}^2$$

$$m_1*v_{1i} = m_2*v_{2i}$$

It is these physics laws which are the foundation of many considerations from the theoretical to the practical and are used to address ideas concerning the generation of energy, its transfer, and use.


**Purpose :** To compare the Gravitational Potential Energy of a spherical mass
( ball ) on a ramp to its Kinetic Energy.

**Purpose :** To measure the Kinetic Energy of a ball before it collides with another
stationary and equally-massed ball and then measure and compare
the Kinetic Energy of the impacted and now in motion second ball to
the first ball.

**Purpose :** To measure and compare the Momenta two different spherical balls.
The initial momentum from the impacting ball as compared to the
momentum of the impacted ball.


**Note :** There are 2 materials lists depending on what you have and want to use.
They both follow the same procedure for the Activity and use the same
formulae.

## Materials ( Basic Materials ) :

- 2 Similar Size & Mass Marbles,
- 4 dowel rods ( each ¼ in. dia. and about 1 yd. long ),
- 2 Stopwatches,
- Mass Scale,
- Ruler,
- Clear Tape,
- Thin book or an item to create an elevation with,
- Level floor area,
- Slide Rule

## Materials ( Larger Scale Materials ) :

- 2 Billiard Balls,
- 2 dowel rods ( each ¼ in. dia. and about 1 yd. long ),
- 2 PVC pipes ( each ½ in. dia. And about 1 yd. long ),
- Mass Scale,
- 2 Stopwatches,
- Ruler,
- Clear Tape,
- Thin book or an item to create an elevation with,
- Level floor area,
- Slide Rule

## Procedure :

1) For either set of materials, the goal is to set up a shallow-sloping hill ( either two dowel rods or two PVC rods atop a short item ( book )) that leads into

2) Two dowel rods that become a track for the ball to go down the slope and roll onto the level track.

3) Note : You will have to work on this to make it run correctly and may need tape so that the rods do not become separated with the ball on them.

4) Note : It may be useful to use an item that accomplishes two tasks – A) It holds the track in place and B) allows for the ball to roll along it unaffected. Coasters may help in this.

5) Measure the mass of each of the balls B1 ( impactor ) (M1 ) and B2 ( impacted ) ( M2 ).

6) With the track in place, place one of the balls on the level track at about the midpoint of the track.

7) Use the ruler to find two stretches of track that are each 20 cm before the ball ( up to impact with it ) and 20 cm after the front edge of the ball.

8) Be sure your stopwatches are set at zero.

9) Have several practice runs to become efficient at timing.

10) Start the impacting ball on the track. Note : wherever you determine its starting height, this will not only remain the same for all trials, but be sure to measure this height with respect to the midpoint of the ball on the hill and once it is on the level track. Record this value ( h ).

11) While practice timing, test the run of the track to see that the impactor ball rolls down the hill onto the level track and impacts the other ball. Have the angle shallow enough so that you can start and stop the timers as needed. Also note if the impactor ball continues to roll or not. Try to find a shallow enough angle and short enough run so that it impacts the other ball and essentially comes to a stop and the other now has momentum.

12) Once the system is operational, conduct 3 trial runs where you measure the distance of the run ( D1 & D2 which are equal ) and the time for the runs ( $t_1$ & $t_2$ ).

13) Be sure to record both the times ( $t_1$ & $t_2$ ) for the Impactor and Impacted Balls. This is probably the most critical value in the exercise.

14) Calculations :

15) Compute an Average Time for the Runs ( $t_{1Ave}$ & $t_{2ave}$ )

16) Determine the Average Speed of each of the balls ( $v_{1Ave}$ & $v_{2Ave}$ )

17) Calculate the Gravitational Potential Energy of the Impacting Ball ( PE )

18) Calculate the Kinetic Energy of the Impacting Ball ( KE )

19) Compare the gravitational potential energy to the kinetic energy for the impacting ball. ( Treat the grav. PE as if it were the ideal value and the KE as the measured value and find the percent 'lost' or difference as it were in this case ).

20) Calculate the Impacted Ball Kinetic Energy ( $KE_2$ )

21) Compare the kinetic energy of the impacting ball to the kinetic energy of the impacted ball. Again treat the impactor ball KE as if it were the ideal value and the impacted ball KE value as if it were the measured and compute the percent 'lost'.

22) Calculate the momentum of the impactor ball ( $p_1$ ) and the momentum of the impacted ball ( $p_2$ ) and compare these results in a similar fashion where $p_1$ is treated as the ideal value and $p_2$ as the measured value.

23) Note for all of the aforementioned comparisons, in most cases the initial values tend to be higher than the final values as this is not an ideal case and some of the energy and momentum are 'lost'. Since energy cannot be created nor destroyed, the question is : where did it go?

## Data :

B1 : M1 : _____ ( kg )
B2 : M2 : _____ ( kg )

Height of Impactor Ball on Hill :

h : _____ ( m )

Distance Traveled by Impactor and Impacted Balls :
    Note : Values should be the same

D1 = D2 = _____ ( m )

| Trial | Time ( s ) [ $t_1$ ] | Time (s ) [ $t_2$ ] |
|---------|----------------------|---------------------|
| 1 | | |
| 2 | | |
| 3 | | |
| Average | | |

## Calculations :

Be sure to use your Slide Rule!

## Average :

$$\text{Average Value} = \frac{\Sigma x}{n}$$

> Note: 'x' is the quantity in question and 'n' is the number of values measured.

## Speed :

$$v = \frac{d}{t}$$

## Gravitational Potential Energy :

**PE = m*g*h**

> Note : g is taken as 9.8 m/s$^2$

## Kinetic Energy :

**KE = ½*m*v$^2$**

## Momentum ( magnitude ) :

**p = m*v**

## Conservation of Energy :

**E$_{Before}$ = E$_{After}$**

Here :
> ( Gravitational Potential Energy of the 1$^{st}$ ball =
> Kinetic Energy of the 2$^{nd}$ ball )

> **PE = KE**

## Conservation of Momentum :

**p$_i$ = p$_f$**

Here : **Impactor Momentum ( p$_1$ ) = Impacted Momentum ( p$_2$ )**

## Percent Difference from Ideal :

$$\%D = \frac{[\text{ Ideal-Measured }]}{\text{Ideal}} * 100\%$$

## Conclusion :

What sort of conclusions can you draw concerning your investigation?
Were the initial values nearly equal ( ideally it would be so ) to your final
values for energy and momentum? With some difference, where did that
energy go? ( consider such things as heat, sound, et al )

# Activity #26
## Doll Bungee Jump Activity
Grade Level : High School
Math Level : Calculating

At one point or another most of us have heard of and seen bungee jumping. Safety is paramount in such an endeavor. This Activity does not involve people, but creates a small-scale version where a doll, such as Barbie or GI Joe or any other similar size doll, can be a bungee jumper for math and science considerations.

Here we are going to use rubber bands which stretch in a limited range in a manner akin to Hooke's Law ( see that Activity for more on this ). In actuality rubber bands only exhibit this over a short range of stretch and at longer distances is shows what is referred to as hysteresis ( a non-linear relation ). Here since the masses are small and the number of rubber bands relatively large the linear relation is reasonably accurate.

The basic goal is to obtain data in the lab ( so to speak ) on a small scale which will be used to create a line equation to determine the number of needed rubber bands connected to each other in a slip knot fashion so that when the doll is dropped from a higher chosen height there is a balance of safety versus the maximum thrill of the fall ( that is to say the doll comes as close to the ground as possible but not hitting it and causing injury ).

For those interested in the math of the Activity this one requires gathering data, data analysis by graphing data points, finding the slope and intercept of a line and use the line to find a reasonable estimate for the needed number of rubber bands to maximize thrill versus safety for the doll. It should be found that the distance the doll will fall to has a directly proportional relationship to the number of rubber bands employed in a given trial. So it can be said that this exercise ( good for math and/or science ) is an examination of a linear function.

**Purpose :** To determine the length of bungee cord needed as determined from graphed data to reach certain maximum yet safe lengths of stretching potential

**Materials :**

- Rubber Bands ( best to have uniform type bought in large amount – typically bagged or boxed and found in an office supply store ),
- Measuring Tape,
- Stairway with railing ( so as to provide the maximum vertical distance safely ),
- Doll ( Barbie or GI JOE – a standard type doll about 7" works well ),
- Table,
- Stack of books or a plastic crate ( best choice ),
- Plastic cutting board or rod ( acts as jumping off point from the crate atop the table ( all held by C-clamps )),
- C-clamps ( 2-3 ),
- Goggles,
- Graph Paper,
- Slide Rule

Note : The testing of both the short falls with increasing numbers of rubber bands to generate data points and the final long fall testing the estimated number of rubber bands needed should be done with parental permission and supervision. Note – wear your goggles and be careful of the rebounding doll.

Note : In the case of the maximum fall, this is an agreed to distance ahead of the experiment and must have parental permission and supervision. It is also recommended to be conducted by the parent(s) as needed. Never engage in unsafe conduct involving vertical heights, stairways, railings, and the like.

Note : The number of rubber bands depends on the chosen maximum falling distance the bungee-jumping doll will fall. Note that for the trial run to gather data one will need initially up to 15-20 ( probably less depending on size ) rubber bands and then you will determine from your generated equation from the data points the estimated number needed to safely fall and return from the chosen maximum height ( this could be a large number – 30 and up )

Picture of possible Dolls ( Barbie on left, GI Joe – Jane version on right – note the rubber bands besides them. Also important is the fact that other than a basic outfit, you should have no other loose items on the doll ( shoes, gloves, necklaces, hairpieces, et al ) :

**Procedure :**

1. Set Up for Activity :
2. 
3. First a Maximum Fall Distance place is chosen and known ahead of time. Its height is measuring for the maximum fall. Note that parental permission and supervision is needed here.
4. Read the Materials list and Notes to be sure you have the correct and needed materials. It is important to have parent permission and supervision as well as wearing goggles. Be sure that the doll used has no loose materials associated with it that can fly off when undergoing the bungee jump.
5. The initial set up involves a small scale version of the bungee jump so as to collect data points, generate a graph and then use the equation of the line formed to determine the number of rubber bands needed for a much larger scale bungee jump and then test that estimated value. The goal of the successful bungee jump is to have the doll avoid injury yet at the same time maximize the thrill of the fall ( be as close as possible without contact ).

6. The initial set up involves a stable table where a plastic crate is turned so as to be its tallest in the vertical direction and then it is secured to the table with one or more C-clamps ( Note if it is a typical crate with holes and it is hard to find a place to clamp across, then place a thin book that fits into the crate to act as a flat plane ). Note : In using C-clamps on a table or book first you need parental permission and supervision – next it may be a good idea to use a thin block of wood to be placed between the clamp and the table or the crate so that it prevents damage.

7. Next – at the highest position ( the top of the crate ) now attach the chosen rod or thin board to be used. C-clamps or tape may be used. It needs to be reasonably secure not only that it does not become unsecure, but also so that it does not vibrate too much when the doll is bungee jumping – the greater the vibration, the greater the error in the data.

8. The attached rod / board projects out over the edge of the table so as to act like a diving platform essentially.

9. Now set the measuring tape on the floor and extend it up to the platform and lock it in place there so that measurements can be taken against it.

10. A critical idea here is that even in executing the trial runs it will require help. It is best to have 2 people to do the experiment. One will hold and then let the doll fall while the other is positioned in such a manner to see the doll fall and be able to watch the maximum extension of the fall of the doll against the measuring tape – and then comes up to the tape and read the value ( to the nearest cm reading ) for each trial.

11.

12. How to Attach Rubber Bands :

13.

14. It is best to use what is referred to as a double-loop when not only adding rubber bands in sequence ( two per each trial is recommended ) but also when first securing the dolls feet.

15. The double-loop is made when there are two rubber bands and one loop is wrapped over the other stationary band and one edge of the looping band is drawn back through the band's other side after wrapping around the stationary band. This is what is called a slip knot. Research it if further information is needed.

16. With each trial add 2 ( be consistent for each trial ). Always attach at the end of the jump line that is attached to the diving platform.

17.

18. Activity : Trial Runs for Data

19.

20. At this point you should have all components set to do the trial runs of the bungee jumping doll. As a reminder it is best to have 2 people to do the exercise since one is letting the doll fall and rebound while the other is reading the scale.

21. With each trial ( a set number of rubber bands ) perform it at least 3 separate times and record each of the results.

22. Collect data for at least 3 ( preferably 4 or 5 ) data points. Note, if needed do to running out of room for the fall distance of the doll and the floor then consider using only one rubber band for each trial run.

23.

24. Calculations :

25.

26. Note : do conversions when and where needed ( see Calculation section ).

27. With all the data collected the first calculation is to calculate the average jump distance for each value of rubber bands. Use the Average Formula. Place these values in the second table.

28. Now graph the Number of Rubber Bands ( x-axis ) vs. the Average Jump Distance ( y-axis ) as data points.

29. Using only the data points draw a best fit line. Notice the direction here – Only the data points – your line should NOT go through the origin of the axes but instead should have a y-intercept ( which is important ).

30. For this line determine the Slope by choosing two points on it and using the slope formula.

31. Now using the Line Equation, plug in the value of the slope and one of the points on the line and determine the y-intercept ( b ).

32. Compare your intercept value to the measured height of the doll. Use the percent error formula to determine the % Error.

33. Now With the line equation completed now let y equal your maximum distance of fall and determine the needed number of rubber bands ( x ). With parental permission and supervision test this hypothesis to see if it works.

34. Did your prediction work – if not, what things are you potentially overlooking. Try the experiment again after first going over your data and rechecking the calculations. Also consider such things as the use of the rubber bands – for example there are ones that are attached for some time to the doll while others are not – could they be too stretched out while the new ones are not as flexible – what can be done to change this ( stretch them ahead of attachment ).

## Data :

Chosen Maximum Distance of Fall : _____ m

Measured Height of Doll : _____ cm

| Trial | Number of Rubber Bands in Jump ( n ) [ x ] | Distance of Jump ( cm ) [ y ] | Distance of Jump ( cm ) [ y ] | Distance of Jump ( cm ) [ y ] |
|---|---|---|---|---|
| | | | | |
| | | | | |
| | | | | |
| | | | | |
| | | | | |
| | | | | |

| Number of Rubber Bands in Jump ( n ) [ x ] | Average Distance of Jump ( cm )[ $y_{ave}$ ] |
|---|---|
| | |
| | |
| | |
| | |
| | |
| | |

## Calculations :

Be sure to use your Slide Rule! : )

Slope :

$$m = \frac{\Delta Y}{\Delta X}$$

Average :

$$y_{ave} = \frac{\sum_{i=1}^{n} y_i}{n}$$

Equation of Line :

$$y = mx + b$$

Percent Error :

$$\%E = \frac{[\text{ Actual Measurement } - \text{Predicted Measurement }]}{\text{Actual Measurement}} * 100\%$$

Conversions :

1 m = 100 cm
1 in. = 2.54 cm
1 ft = 12 in.

## Conclusion :

How does your estimated values for the needed number of rubber bands compare to the actual distance – was it safe or not ? What sort of factors could have affected the outcome ? How close was your predicted measurement for the height of the doll as compared to the actual measurement?

# Activity #27
## Centripetal Force Measurement Activity
Grade Level : High School
Math Level : Calculating

Many of us have ridden or driven in some sort of vehicle that has gone around a corner or curve in the road. What seems to be a force that acts on our bodies in the vehicle so that we seem to move as if away from the motion of the vehicle as it turns is not what it seems.

Actually, we have mass, hence we have inertia. While in the vehicle we are moving in a 'straight' line going into the curve as does the vehicle. A force is exerted on the vehicle to steer it into the curve. It is the frictional force between the tires and the road that creates the force to cause this change in motion from a straight line. Our bodies, also in accordance with Newton's Laws of Motion, tend to continue in a straight line motion – so this seems like we are being pushed away from the vehicle initially. The vehicle now rounding the bend in the road is moving toward us and forces us along with it through the curved path. The force we experience by the car that is moving us toward the center is what is called a Centripetal Force.

Defined, a Centripetal Force is any force that causes an object to follow a circular path and is the term given to any force that acts at right angles to the motion of an object undergoing it. The term centripetal means 'center-seeking' or 'towards the center'.

Centripetal Forces are quite common in Science and are found in many areas. On an atomic level consider an electron in motion towards a massive nucleus. Close enough the electromagnetic forces on the electron cause its initially straight line motion to become the orbital path about the nucleus. In a somewhat similar manner the Moon is moving through space, yet the Gravitational Force between it and the Earth cause it to move in a somewhat circular path about the Earth ( more technically each moves about a common center of gravity between them that is about 1,000 miles deep in the Earth ). This idea then extends to the Planets in orbit about the Sun where they have elliptical paths ( near circular ).

We take advantage of Centripetal Forces. The spinning tub in a washer at a very high rate exerts centripetal forces on the wet clothes, while the holes in the tub allow the water to escape. The water does not have the centripetal force acting on it, it is merely following the motion already imparted to it and moves in a straight-line path. Another device is the centrifuge used in chemistry to separate solutions into components of differing molecular weights.

Why then do we have the misconception of a force acting away from the center, or what is called a centrifugal force? It is because of our sensation of inertia when undergoing a circular motion. To demonstrate it consider a mass at the end of a string being whirled around ( do this mentally and not physically ). If the string holding it and acting as the centripetal force were to snap, the mass would not fly directly away from the center of the path, but instead would move off at a tangent to the circular path ( once again, as noted in Newton's Laws of motion ). Now imagine a sensor on the mass measuring force. The force would not only have magnitude ( the amount ) but also direction which would be back along the string.

This idea of a centrifugal force however does have a potential application. Imagine you have a circular spacecraft that rotates about a central hub. Astronauts could stand inside the outer ring in the spacecraft where the centripetal force of the craft makes it seem that there is a centrifugal force that acts like weight. To illustrate this notion, see the space city in 2001 A Space Odyssey movie.

Our Activity requires parental permission and supervision since we are going to have a whirling mass ( a small washer or nut ) on the end of a string that goes through a spool. Why not just hold it? We need to have a way to measure the force countering the centripetal force, so we will use a mass hanging vertically to oppose the horizontally moving mass. The vertically hanging mass has gravitational force acting on it and this is in balance with the centripetal or horizontal force acting on the whirling mass when it is balanced.

From this we can have a predicted value for the speed it should be moving and from our measures of motion ( rotations and time ) we will compare these values.

( Note : For more information on Forces, see Newton's 2$^{nd}$ Law Activity either online at www.cosmicquestthinker.com or in volume one of The Inquisitive Pioneer ).

**Purpose :** To Measure the Rotational Speed of a Mass where its Centripetal Force balances the Gravitational Force acting on a supporting mass and compare the Predicted speed to the Measured Speed for such a system

## Materials :

- Spool,
- Kite String,
- Measuring Tape,
- Small Nut or Washer ( ¼" at most ),
- Bolt with Nuts or Washers ( 2"-3" bolt or about 10 washers ),
- Mass Scale,
- Stop Watch,
- Goggles,
- Slide Rule

## Photo of Most Experiment Materials :

Note : Safety Goggles not in photo, but is a necessity!

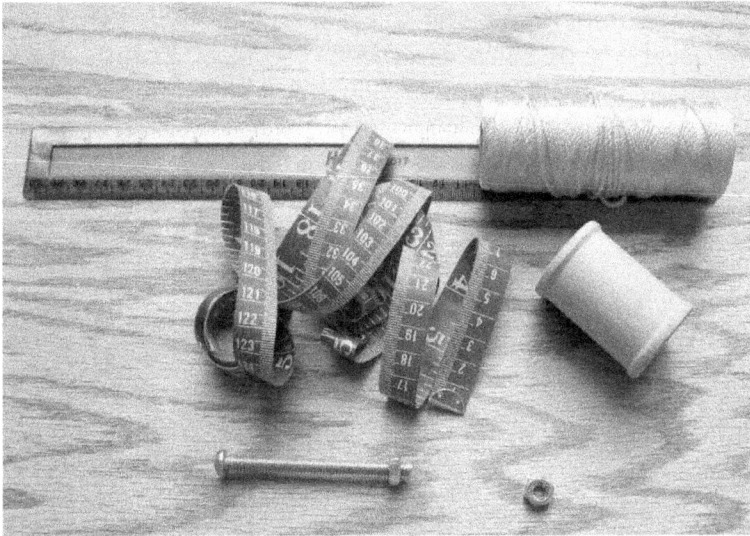

Arrangement of system when assembled :

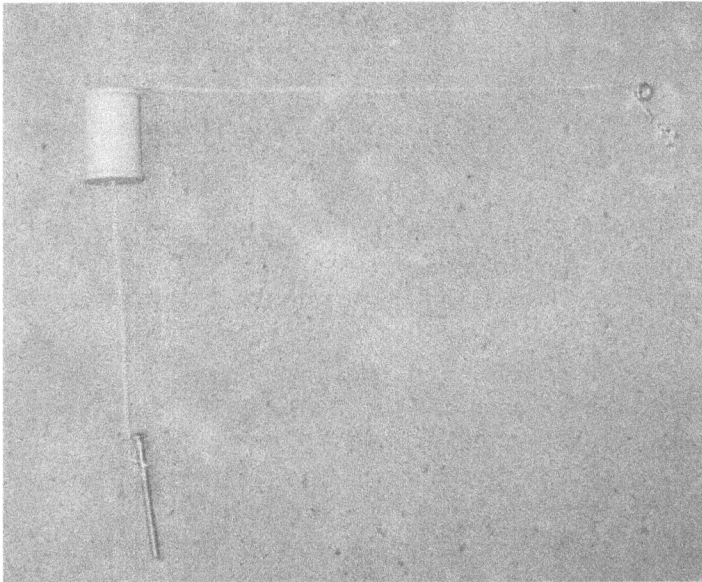

**NOTE :** This Activity requires parental permission and supervision. Also you must wear goggles through the entire experiment – which must be done by all participants. It is best to have parents read and perhaps do the motion aspects of the lab. All must work safely.

**NOTE :** This Activity uses a string attached through a spool at both ends where the smaller mass ( one nut or one washer ) is swung in a circular manner horizontally above one's head while the other mass hangs vertically down from the spool. Since there is something in motion the person whirling the mass must be sure to be aware of others, of objects while engaged in the motion of the mass. The maximum safety is where this is done outside. All participants should have goggles, stand in a place where they are at maximum safety ( Interestingly if there are 2 people, they should stand close to each other since the mass on the end of the string will be above both of them and would have the least chance of hitting anyone – also each has a job – one spins and counts the number of revolutions of the mass while the other operates the timer ).

## Procedure :

1. Be sure to read the preceding Notes where safety is discussed. You cannot do this Activity without parental permission and supervision. At all times in the Activity keep your goggles on, be aware of others, and be safe in your actions.
2. You can use the suggested numbering system for trials and subsets of trials or use your own.
3. Determine which mass will be the Horizontally Spun Mass ( should be no more than 1 small or medium sized nut or 1 or two washers ). Measure and record the mass for the data table ( m ). Note, in subsequent trials this mass will remain the same.
4. Determine which masses will be the Vertically Hanging Masses ( best to use a bolt to which nuts can be attached ). Measure and record the mass for the given Trial for the data table ( M ). Note in subsequent trials, this mass will vary.
5. Cut a piece of string ( between 60 cm to 100 cm at most ) and string it through the spool ( note it is best to have as large a spool as possible to make it easier to hold and it should be either wood or plastic ).
6. To one end of the string securely attach the Horizontally Spun Mass and do the same for the Vertically Hanging Mass ( see photo ).
7. It is best to do this outside and have no one nearby.
8. The following is for the one who is doing the spinning of the mass. Be sure to note the recommendations, where parents are involved in this process. Before the actual measurements of the spin rate, it is best to practice your technique at this so as to work out any problems. Hold the spool and vertically hanging mass in your hand leaving about 30 cm for the horizontally spun mass. Begin a slowly increasing spin so that it is above your head and so that the mass going around is parallel to the ground. Develop a spin rate that is controllable, countable. You are doing well when the hanging mass is free and there is some string between it and the spool.
9. With step 8 in place, now do it where you are taking measurements. As noted it is best to have the person timing the event as near the person whirling the mass as possible.
10. Once the spinning motion is steady, begin the timer. Count a given chosen number of complete rotations ( can be 5, 10, or whatever you choose ). Record this value ( n ) and record the amount of time ( t ) it takes to complete this number of rotations.

11. Once completed as you stop the spinning mass hold the string on the vertical hanging mass side just where it enters the spool so as to hold it in place. You will have to measure the radius of arc ( R ) that the horizontally spun mass is traveling. Record the radius ( R ) in each trial.

12. Do steps 10 and 11 at least 3 times since you are going to determine the average values of both 'n' and 't'. Note, you should choose the same 'n' each time. You will also measure and average the radius of arc ( R ) traveled.

13. At this stage, one trial is complete and you can test other vertically hanging masses ( M ) in further trials. Note that each is treated separately when performing the calculations, though the results may be compared.

14. Calculations :

15. As noted you have to average the number of rotations ( n ), the amount of time for the rotations ( t ), and the radius of arc traveled ( R ).

16. For each trial separately use the average values in calculating the predicted tangential speed ( v ).

17. Using the average values, compute the average angular speed ( $\omega$ ) and then the measured tangential speed ( v ).

18. Compare results of the predicted versus the measured and consider why they are not exactly the same.

19. If you have done further studies of other vertically hanging masses and have done their computations, is there a pattern – for example, as the vertical hanging mass goes up, does the speed increase or decrease? To determine this more accurately, create a graph of this of Tangential Speed measured on the y-axis and the Increasing Vertically Hanging Mass values on the x-axis. Try a best fit line and determine slope. If it appears to curve, try a log-log plot of the values and take the slope of a best fit line to see if there is a power relation at work here.

20. Note : If interested you can independently compute the horizontal and vertical forces using the formulae provided for your given trials as well.

## Data :

Trial # _____
       ( Use Roman Numerals : I, II, III … )

Horizontally Spun Mass [ m ] : _____ g

Vertically Hanging Mass [ M ] : _____ g

Radius for a Given subset of a Trial

| Subset of a Given Trial [ numbered 1,2,… ] | Radius Measure [ R ] |
|---|---|
| I1 | |
| I2 | |
| I3 | |

Amount of Time for One Revolution :

| Subset of a Given Trial [ numbered 1,2,… ] | Number of Revolutions [ n ] | Amount of Time for 'n' Revolutions [ t ] |
|---|---|---|
| I1 | | |
| I2 | | |
| I3 | | |

Note : It is best to use whole numbers for 'n', such as 5, 7 or 10 for example

## Calculations :

Be sure to use your Slide Rule!

Average :

$$\mathbf{X_{ave}} = \frac{\Sigma x}{N}$$

( x = a given value in the set, N = the number of elements in the set, Σ means the sum of the items )
( Here the average is the Average Speed of Revolution, so it is best to let both 'n' and 't' be 'x' independently and determine the average for each trial and then take the ratio of '$n_{ave}$' and '$t_{ave}$' in the angular speed formula )
( Note also to determine the average radius in a given trial )

Vertical Force ( Weight ) :

**F = M*g**

( M = vertically hanging bolt and nut mass,
g = acceleration due to gravity taken as 9.8 m/s/s )

Horizontal Force ( Centripetal ) :

$$F = \frac{m*v^2}{R}$$

( m = mass spun horizontally, v = tangential speed,
R = radial distance of mass from center of rotation )

Predicted Tangential Speed : ( Where Forces are Balanced ) :

$$F_v = F_H$$

$$M*g = \frac{m*v^2}{R}$$

$$v = (( M*g*R ) / m )^{1/2}$$

Tangential & Angular Speed Relation ( Measured Speed ) :

**v = ω*R**

Angular Speed :

$$\omega = \frac{\Delta\theta}{\Delta t} = \frac{2*\pi*n}{t}$$

## Conclusion :

How close did your predicted speed values compare to the measured
speed values? Why do you think so? As you increased mass, where the
values more similar or less similar? Why do you think so?

# Activity #28
## Air Rocket Launch Analysis Considerations Activity
Grade Level : High School
Math Level : Challenging

This Activity is different from most in one important way – either one must purchase an air or water powered rocket from a retailer or find online instructions to properly create one ( often from a soda bottle ) while still purchasing the needed launch pad elements to complete the rocket system. In any measure it is best to buy the rocket system – rocket and launch pad – for optimal safety. Instruction to assembly are not given here. Due to the type of Activity this is, parental permission, supervision, and participation is a must. Safety at all steps is required.

As this Activity does not look to construction of a rocket and at best only recommends the proper use of an air-powered or water-powered model that one must follow the safety considerations from the manufacturer for, the overall goal and purpose of this Activity is the mathematical analysis of the proper launch of these systems only.

Rockets always have captured our collective imaginations and spirit for soaring adventure. To illustrate the complexity of an issue, the phrase 'it's only rocket science' is used. It was rockets, such as the Saturn V that lifted the Apollo spacecraft to the Moon to usher in humanity's greatest and most distant adventure. Rockets are used in the public for fireworks, the military for weapons, as well as in everyday life to lift all sorts of satellites and spacecraft into space. The satellites are not only the ones that orbit the Earth and relay electromagnetic information such as telephone transmissions, photos, and other images, but also there are ones that rocket off to far away worlds such as Mercury, Mars, Jupiter, and even now Pluto to send scientific instruments to understand these distant worlds. Other satellites examine stars, like the Sun, as well as distant galaxies.

Chemical Rockets are the most common type and the first rocket came from the development of black gunpowder. 9th century Chinese Taoist alchemists discovered this while developing materials in their search for the elixir of life. In time, many early experiments were done such as creating incendiary fire arrows, making bombs, cannons, and rocket-propelled items. The first recorded account of fireworks comes in a statement in 2 Vopiscus,, Carus, Numerianus et Carinus, ch. Xix which notes that fireworks were performed for emperor Carinus ( 282-283 ). A reliable reference to rockets comes from China in the Ko Chieh Ching Yuan ( The Mirror of Research ) which notes that in 998 AD a man named Tang Fu invented a rocket of a new kind, this one having an iron head. There are many other references, though obscure and some not verified, but needless to say that there were a great deal of initial efforts in China. In fact, the news of rockets came to Europe due to the Mongols acquiring the information from the Chinese and even the Mongol leader Genghis Khan used it, such as in the Battle of Mohi in 1241 where the Mongols were fighting the Magyars. In a time period between 1270 and 1280, Hasan al-Rammah wrote al-furusiyyah wa al-manasib al-harbiyya ( The Book of Military Horsemanship and ingenious War Devices ) which included 107 gunpowder recipes where 22 of them are for rockets. The term 'rocket' comes later from the Italian Rocchetta ( i.e. little fuse ) which was a term given by Muratori, an Italian artificier in 1379.

For more than two centuries ( 1600s and 1700s ) the book "Artic Magnae Artilleriae pars prima ( Great Art of Artillery, the First Part aka The Complete Art of Artillery ) by Poilish-Lithuanian Commonwealth nobleman Kazimierz Siemienowicz is used throughout Europe by the Dutch, Germans, French, and English. The book has a set which became the standard designs for rockets, fireballs, other pyrotechnic devices for both military and civil use. It detailed the construction, production, and properties of rockets, even multi-stage rockets and having rockets with delta wing stabilizers.

The first iron-cased rockets were made in 1792 by Hyder Ali and his son Tipu Sultan, rulers of the Kingdom of Mysore in India and used against the larger British East India Company forces in the Angle-Mysore Wars. By now rockets had a range of up to 2 km, which was much further than those of the past. This too lead to the British interest in them in the 1800s. Rockets make appearances more and more often, such as noted by Francis Scott Key 'rocket's red glare' in the 1814 Battle of Baltimore and even in Waterloo against Napoleon. The 1800s saw greater and greater understanding and technology employed so that a rocket would spin when launched to help stabilize it. In time rockets gained in the areas of travel distance, size of payload, better design for stability and less air resistance, and hence gained accuracy.

Rockets come into the public not just from reality but in fiction, such as in the stories of Jules Verne and H.G. Wells. The idea of rockets going into space and to other places began to become an idea. In 1903, high school mathematics teacher Konstantin Tsiolkovsky ( 1857-1935 ) published 'The Exploration of Cosmic Space by Means of Reaction Devices'. It is considered the first serious scientific work on space travel. In it he develops what is now called the Tsiolkovosky rocket equation, which embodies the principle that governs rocket propulsion. He even thought that liquid hydrogen and liquid oxygen would be the best fuels, which are later creations and used in modern NASA rockets. By 1912, Robert Esnault-Pelterie publishes a lecture which independently derives the Tsiolkovsky equation as well as the energy needs to make round trips to the Moon, planets, and the like. He also proposes the use of atomic power ( i.e. radium ) to power the drive system.

Inspired by these ideas as well as the science fiction of the day, Robert Goddard set out to examine the idea of rockets and space travel as early as 1912. He had been inspired by H.G Wells and sued Newton's Principia to explore the notion of rocket flight. He felt that solid-fuel rockets needed three improvements – use a small combustion chamber instead of the whole area, next arrange the rocket in stages, and change the exhaust nozzle to increase the speed of the propellant. By 1920 Goddard had published these and other ideas in A Method of Reaching Extreme Altitudes. In 1926 he launches the first successful liquid-fuel rocket using his design ideas.

Rockets undergo rapid transitions due to World War II and the efforts on both sides, such as the British and Americans on the ally side as opposed to the Nazis in Germany and their infamous V-2 rocket. After the war, the efforts do not slow down, but are for demonstration purposes such as in the Cold War era of Russia and the United States. In fact both their personnel as well as materials came from the Germans. For example there is W. Von Braun who redevelops the V-2 for the Americans into the American Redstone, which is used in the early space program of NASA. By now the world of solid chemical and liquid fuel rockets have been established and during the next three decades ( 1950s to the 1970s ) there is much use and development of scope and scale of rockets and their use.

For example there were launches of the first satellites, such as the Sputnik I October 4, 1957 by USSR and the first person in space Yuri Gagarin aboard Vostok I on April 12, 1961 followed by the US with their first satellite on January 31, 1958 the Explorer 1, and on May 5, 1961 aboard Mercury Freedom 7 is Alan B. Shepard Jr. This back and forth goes on for more than a decade with launches of satellites and people. The Soviets had programs such as Vostok, Soyoz, and Proton, while the Americans had Mercury, Gemini, and Apollo programs. The main race was to the Moon, which is accomplished by the United States on July 20, 1969 aboard the Eagle landing craft coming from the Columbia command module launched atop the Saturn V rocket and having astronauts Neil Armstrong and Ed 'Buzz' Aldrin as the first to set foot on the Moon. The Saturn V is still the largest and most powerful launch vehicle ever used. It stood 363 ft tall, had a mass of 3.7 million pounds, , could carry to the moon a payload of 100,000 lbs, operated by 5 Rocetdyne F-1 engines with a thrust of 7.6 million pounds-force with a burn time of 360 seconds, and propelled by liquid oxygen and liquid hydrogen fuel.

As an aside, rockets, or at least their engines, are not just for launch vehicles. There are rocket cars, sleds, trains, rocket-powered jet packs, rocket-powered rapid escape systems – like ejection seats, along with all the other conventional uses of rockets such as launching satellites, spacecraft, and people. This does not include the fact that model rocketry is an exciting hobby too.

Simply throwing an object, like a baseball, into the air does not make it a rocket. A rocket has its own fuel system to move it. Unlike a bird that uses air and unlike a ball that is given enough kinetic energy due to a throw, the rocket has an internal energy system. The fuel has some sort of potential energy that is translated into the kinetic energy of the rocket itself once activated. The simplest rockets simply used pressurized air or water as the expellant. The most complex use chemicals and rocket engines to have combustion take place, even in the void of space, to create a propellant. Regardless of the complexity, all rockets are propelled by the release of mass through an exhaust portal while the rocket moves, according to Newton's 3rd Law of Action-Reaction Forces in the opposite direction. Rocket engines work by action and reaction. Rockets follow the conservation of energy and conservation of momentum laws. If we were to measure the mass in the exhaust and its velocity and compute it as a product it would match the mass and velocity of the rocket in the opposite direction. This is why a rocket can operate in space. It does not push off of something, it is merely expelling exhaust in sufficient quantity and speed to cause a reaction force that acts on the rocket to move it in the opposite direction. To speed up, fire the rockets in the direction the craft is already moving, to slow down, fire the rockets in the opposite direction of travel for the rocket and to move right, fire the rockets left and vice versa. A rocket, by definition then, is a missile, spacecraft, aircraft or other vehicle which obtains thrust from a rocket engine. In all rockets, the exhaust is formed from propellants carried within the rocket before use.

$$F = \frac{dp}{dt}$$

'dp/dt' is the change of momentum with respect to time. In this case,
mass is no longer a constant, as it is changing, much like the velocity
is too.

One might wonder why not just use rocket fuel and have most things rocket powered? Simple, cost and effectiveness. Rockets are noisy, have large amounts of heat, flame, and the like. Some of this is lost energy and does not go into the actual overall propulsion of the system. The energy density of rocket propellant is 1/3 that of conventional hydrocarbon fuels and the bulk of the mass in the rocket is the oxidizer. From the equations, only when high speeds are needed are rockets effective. To illustrate efficiency, the shuttle turns out to be about 16% since the KE of the shuttle is about 3TJ ( a 100,000 kg vehicle moving at an altitude of 111 km at 30,000 km/hr ) while the chemical potential energy of the fuel is 200 TJ.

The forces affecting a rocket are : Thrust as supplied by the engine, gravity acting towards the Earth, Drag which acts opposite to motion and comes from the atmosphere, and Lift. The main force that all rockets go against is the force due to gravity first and secondly air resistance. When we toss a ball in the air, it decelerates at 9.8 m/s/s for each moment of its time of travel. It has zero speed at the top of its flight path, and then returns to the Earth, accelerating at the same rate as deceleration. If it returns to its starting point, it will have the same speed it left with, only now moving in the opposite direction ( assuming no air resistance ). Interestingly, the rocket does the same thing. The only change here is when the fuel system of the rocket is active, it will have a net acceleration ( much like the accelerated ball to be thrown ) countering acceleration due to gravity. In the ideal model situation, the rocket though returns to the Earth at the speed it left and like the ball, at its highest point has zero speed.

Our first Rocket we will consider is only wind powered, but it none-the-less is a rocket and it too is propelled into the sky. We will not consider the use of a chemically-powered rocket in any case at all. But how high does the rocket go? In our Activity, we can use two different methods to determine this value and compare the results. The first is easy to consider. Simply measure the amount of time the rocket takes to launch from the pad and then return to the Earth. Since the amount of time going up equals the amount of time falling down, we take the total time and divide it in half. Taking this value we can quickly multiply by 10 ( rounding acceleration due to gravity value of 9.8 m/s/s ) and find the launch speed roughly. For the altitude, we must do a bit more calculation. To estimate altitude, square the time value ( the half one and not the total trip time ) and multiply by 5 (0.5*10 – again using an estimate for 'g'). In the Activity, use your slide rule to verify your mental calculation estimates, but then use the accepted value of acceleration due to gravity of 9.8 m/s/s to find more precise values.

The second method requires good eye-hand coordination. At the launch of the rocket, you point a device sometimes called an Altimeter ( or Astrolabe ) at the rocket and follow it to its highest point. As you look along the Altimeter through the straw, a string hangs down with a weight on it and measures an angle. At its highest point place your finger on the string to hold it in place and measure the angle. With a known distance from the rocket launch pad to where you are standing and the angle, you can now find the altitude to which the rocket raced aloft.

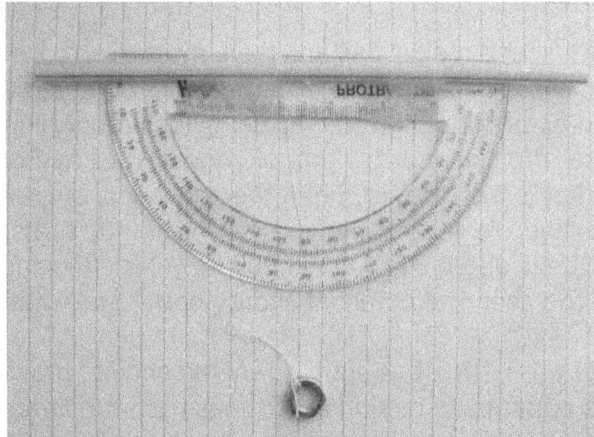

Notice from either consideration, mass is not factored in. Mass matters when one has to determine the amount of fuel needed to lift the rocket and do some given task, such as leaving Earth's atmosphere and heading to space ( the rocket has to achieve escape velocity of 7 mi/s or about 25,000 mph ). In fact most rockets to do this need about 3-4 lbs of fuel for each pound of payload on the rocket.

Another possible Activity can consider launching the wind or water powered rocket at an angle that is measured (this critically assumes we have a launch pad that can safely be angled in a secure fashion – some purchased models have this option - if it cannot, then do not do this).

What if, however we can angle the rocket and the rocket is then shot like a projectile, at an angle instead of straight up in the air? If we know of a rocket that when launched can achieve a certain height and from that we can determine its launch speed, we can then use 2-dimensional vector analysis to separate the horizontal and vertical components of the speed of the rocket. Rockets at an angle are a great way to investigate the behavior of velocity as affected by acceleration due to gravity. We will use the same conclusions that Galileo had come to - that the horizontal speed component is independent of gravity, while the vertical component is not.

Imagine the vertical initial speed is now the hypotenuse of a triangle. The legs of the triangle are the horizontal speed and vertical speed components. The angle that the rocket is launched at can be used with the hypotenuse to determine the component speeds. The horizontal component will remain constant throughout the flight as it is not affected by the acceleration due to gravity, which operates in a vertical direction. The vertical speed will first decrease to zero and increase back to its original value ( assuming the rocket lands on a plane at the same elevation of launch and having to air resistance ).

Picture of v, vx, vy

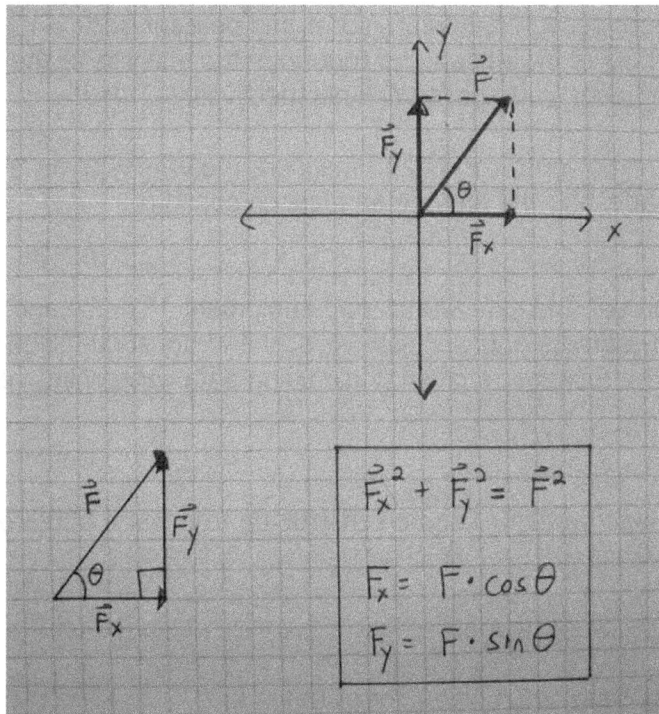

Here let F be velocity and :

$$v_x = v*cos\Theta$$

$$v_y = v*sin\Theta$$

Using the vertical speed, we can then determine the maximum amount of time the rocket will spend in the air. Since the horizontal component of the speed remains constant throughout the trip, we can multiply this determined amount of time by the speed and find the Range, or distance, the rocket will travel horizontally from the launch pad when launched at a given angle. We could also use the Range formula ( as well as height formula ) in Physics for 2-dimensional situations as well to calculate an estimate of these values. The Range we can readily compare to the actual outcome.

An Alternative second Activity can be the construction of a Water-Propelled Rocket, which can be used much in the same manner as the Air-Powered Rocket. In either case, these rockets can be constructed ( instructions not provided here ) but both are readily found from online retailers.

## Activity : Air-Powered or Water-Powered Rocket

**Purpose :** To calculate altitude achieved by an air-powered rocket by two means : timed measurements of ascent and descent and measured angle of inclination and trigonometric analysis of the triangle from launch and observer's perspectives.

**Purpose :** To use an air-powered rocket and measure the effects of air pressure in affecting the altitude and launch speed of the rocket.

**Purpose :** ( If using proper and safe launch pad equipment ) To use and air-powered rocket and estimate and calculate vertical and horizontal speeds as well as range for the rocket at a given angle using data concerning vertical launches

## Materials :

- Air-powered rocket system ( can be constructed from soda bottle or a purchased model – recommended ),
- Goggles,
- Air Pump with pressure gauge,
- Timer,
- Long Measuring Tape or Trundle ( for distance ),
- Altimeter ( purchased or constructed * ),
- Launch Pad ( note – buy one for safety and there are ones that will allow for the angled launch purpose if desired ),
- Protractor,
- Slide Rule

## Notes :
1) The * 'constructed' altimeter occurs in the Altitude, Height, and Distance Activity – see that Activity for materials and construction of the altimeter. ( Need Protractor, straw, tape, string, metal nut ) Also there are these type of items at science supply websites, like Arbor Scientific – theirs is called an Altitude Finder.
2) Safety is always a must. With more than one person in this Activity, others must maintain a safe distance from the rocket when prepared for launch, which includes the one pressurizing the rocket. Adult supervision is recommended. In the process of launching, wearing goggles is a recommended idea. Do not overpressure the system. Be sure that the rocket is not pointed in a manner to strike individuals upon launch.
3) Though one type of rocket is recommended, the air-type due to overall safety and ability to angle it, there are others ( such as the water-powered one ) on the market – the key is to be able to vary the pressure and have a means to read the pressure for the rocket. Be sure to read all safety precautions with each system employed.
4) For most rocket systems, a wide open grassy area is best since hard surfaces can damage the rocket and trees could be an obstacle.

## Procedure :

1) For whatever air-propelled rocket you are using be sure to follow all of the assembly and safety directions.

2) Some of the important assembly directions include the use of goggles, standing as far as is possible for safe launches, securing the launch pad, not pressurizing the rocket with anyone standing over the rocket.

3) Another important issue is to find an area for launch that will allow easy access to the rocket both for vertical launches and angles launches. Be wary of homes, trees, et al.

4) The larger the number of people involved the better, since someone must launch the rocket while another starts and stops the timer.

5) Most often the timer person is standing away from the rocket and will also measure the angle of altitude with the altimeter, so timing of watching the rocket and starting and stopping the stopwatch is critical.

6) The best minimum number of people is 3, since one is the launcher, a second is a timer and a third is the altimeter operator.

7) Note that the fewer the altimeter people, the less accurate the readings, since it is best to have >=2 of these people and average their results.

8) Whatever the number of altimeter readings per trial, average the values of the angle of altitude achieved.

9) If only 1 person, then try to be as accurate as possible for conditions of launch for each launch as possible and average these results.

10) If you have 5 people on the altimeter readings, then you can eliminate the largest and smallest angle measures and average the three middle values.

11) The first set of launches is for vertical readings. Use a set air pressure as prescribed by the directions for the item you are using ( be safe and do not exceed these guidelines ). Record the value used.

12) Perform at least 3 launches measuring the angle of altitude and the time for the entire flight ( launch to land ). Record these values.

13) After calculating the averages for the time and the angle, use these values with the appropriate formulae to determine height ( note there is one for the angle and one for the time value ).

14) Compare and contrast the values of altitude from the angle method and the timing method to see how similar or dissimilar they are.

15) Note that launch speed calculation comes from the time value.

16) For further investigation, one could try other air pressures ( within the acceptable range of air pressures for the rocket as found in the directions ) for comparison.

17)

18) Rocket launched as Projectile :

19)

20) With the appropriate equipment and safety employed, the air-powered rocket launcher can be used to launch the rocket at a given angle.

21) Note, we will use the data for launch speeds as determined in the vertical launch sequence noted above.
22) Following the directions of the manufacturer for launching at an angle ( particularly no one in front of the rocket at any time ) choose a set of angles to use for your considerations.
23) In each of the trials for a given angle, measure the total amount of time the rocket is in flight and measure the distance the rocket travels down range from the launch point. ( Be careful to watch for bounces that the rocket takes, one wants to measure from the initial impact spot! )
24) Be sure to use the same pressure values as noted in the vertical trials as well as the determined launch speeds for these trials. These values are needed to make predictions as to what the range should be.
25) Note : For safety reasons, all observers need to be to the side away from the area where the rocket flies.
26) With the measurements taken, do the calculations for the expected time of flight and predicted range as noted by your calculations by using the former data for launch speed for a given air pressure.
27) You can also do other calculations, such as the predicted height that the projectile rocket went to and range directly from the launch speed without the component method approach if desired.

## Data :

## Rocket as Straight-up launch vehicle :

| Trials | Air Pressure at Launch ( lbs/sq.in.) | Total Trip Time ( s ) | Average Angle ( Θ ) from Altimeter |
|---|---|---|---|
| 1 | | | |
| 2 | | | |
| 3 | | | |
| Ave. | | | |

Note : Average Angle is the average of 3 or more readings by individuals using the altimeter to determine the rocket's maximum height.

Note : The Data above is used to determine both the maximum height of the rocket ( two different ways ) and for determining the maximum initial launch speed of the rocket.

This Second Data Set is the determination of the height from the prior data set in the first table. On the second table calculate the launch speed.

Note : There is only one calculated value from each of the trials as they are all averaged into a single value.

| Max. Height from Time Measures (h) | Max. Height from Angle Measures ( $h_r$ ) |
|---|---|
| | |

| Launch Speed ( m/s ) |
|---|
| |

## Rocket as a Projectile :

This Third Data Set is a set of trials when launching the rocket using at a given a given Air Pressure, and at a given Angle of Launch. The first data table is the Actual Measured Results, while the second is the calculated values from previous data tables.

Angle of Launch : _____

| Trials | Measured Range ( m ) | Measured Time of Flight (s ) |
|---|---|---|
| 1 | | |
| 2 | | |
| 3 | | |

## Calculated Values Table

| Trials | Given v ( m/s ) | Calculated $v_y$ ( m/s ) | Calculated $v_x$ ( m/s ) | Calculated time $t_\Theta$ (s ) | Calculated Range [ R ] ( m ) |
|---|---|---|---|---|---|
| 1 | | | | | |
| 2 | | | | | |
| 3 | | | | | |

Note : 'v' comes from the calculated launch speeds above in the prior data set trials

## Calculations :

Be sure to use your Slide Rule!

For estimated calculations, use acceleration due to gravity as 10 m/s/s

For more accurate calculations, use
Acceleration due to Gravity as 9.8 m/s/s

Arithmetic Mean or Average Value :
For 'n' readings :

$$x = \frac{\sum_{i=1}^{n} x_i}{n}$$

Example Calculation : There should be 3 simultaneous measurements minimally ( can have as many as 5 ) of the rocket's altitude as represented by the angle measured by the altimeter.

Assumption :
  *Time Up = Time Down*

Time to Reach Maximum Altitude Vertically :

$$t = \frac{\text{Total Trip Time}}{2}$$

Launch Speed : ( 't' is the time to reach max altitude value )

$$v = g*t$$

Height from Launch Speed Calculation :

$$h = \frac{v^2}{2*g}$$

Altimeter Measurement of Maximum Altitude of Rocket
  Note : standing at distance 'd' from launch
  $\Theta_{\square}$ is angle measured by altimeter
  $h_{\square}$ is the altitude of the rocket

$$h_{\Theta} = d*\tan(\Theta_A)$$

<u>Vector components of speed :</u>
  ( Θ is the angle launched at with respect to the ground )

**Horizontal Speed : $v_x = v*\cos\Theta$**

**Vertical Speed : $v_y = v*\sin\Theta$**

<u>Time Spent in the Air when Launched at an Angle Θ :</u>

$$t_\Theta = \frac{2*v_y}{g}$$

<u>Range of Rocket when Launched at Angle Θ :</u>

$$R = v_x*t_\Theta$$

<u>Percent Error : Actual Range Distance versus Predicted Range Distance</u>

$$\%E = \frac{[\text{Predicted Range-Actual Range}]}{\text{Actual Range}}*100\%$$

<u>Other Formulae for Determining Range and Height of the Rocket :</u>
  'v' here is the launch speed

$$R = \frac{v^2*\sin(2\Theta)}{g}$$

$$h = \frac{v^2*\sin^2(\Theta)}{2*g}$$

## Conclusion :

The real test is how close do your estimated and more importantly, your calculated values from predictions for outcomes match the actual results. What factor(s) do you think affect the outcome?

Other Calculations :

Aside : Can knowing the gauge pressure and the volume of the cylinder of the rocket provide the potential energy of the rocket and can this be compared to the gravitational potential energy achieved by the rocket?

Summary :

Extension ( or Alternative Variation ) Activity :
    A teacher could have larger groups with this project instead of just
    2 or 3 students and a single rocket. With a handful of rockets and
    a few air pumps, a teacher can assign students to various
    subgroups working with one of the rockets. Note that each group
    can have 2 or 3 students, and that students with proper
    coordination on the part of the teacher, can be in more than one
    group. The Main Launch Group has responsibility to prepare,
    launch, and recover the rocket. Also they have the set up of
    information – the nozzle being used, the pressure launched at,
    et al. The Time Group measures the amount of air time and
    performs both estimated and directly calculated values for altitude.
    The Altimeter Group uses the Altimeter to determine the measured
    height of the rocket and perform these calculations. The Time
    Group and the Altimeter Group compare and contrast results. The
    Engineering Group performs the estimated calculations for range
    of the rocket at a given angle.

# Activity #29
## Lever Exploration Activity
Grade Level : Middle School
Math Level : Calculating


The simple machines are : the pulley, wheel & axle, screw, lever, wedge, and the incline plane. They are a somewhat overlooked item at many levels in all of human technology, from all massive structures to even microscopic-sized-systems today, which are being made to operate as levers, and the like. In our everyday lives, they are found in numerous home tools : scissors, can openers, steps, screws, knives, placement of door handles, all hand tools – pliers, screwdrivers, et al, parts of bikes, parts of cars, parts of engines, and so on.

Remarkably, all simple machines have the same basic outcome. Each of the simple machines can only do one or both of the following : Either change the direction of the force and/or the amount of force. They cannot change the amount of work that needs to be done, but can affect the factors involved in work itself.

One of the most common simple machines ( and being explored in this Activity ) is the lever. It is an 'arm' that rotates about a fulcrum and ,typically, the 'arm' that rotates in the opposite direction is used to first change the direction of the effort force exerted on the first arm and cause a resistance force to be exerted against a load.

The length of the lever arms will affect the multiplying affect on the force exerted. Take for example a nut cracker. One exerts a force at a distance from the fulcrum, while the nut is placed between the fulcrum and the effort force placement, hence has a shorter distance. Let's say that it were exactly at half the distance one is exerting the effort force. From the idea of conservation of energy, the amount of energy going in must be equal to that coming out, and this is measured by Force times Distance. With resistance distance being half ( in the example ) means that the resistance force is twice that of the effort force. The shorter the resistance distance the greater the multiplying factor for the force, hence the amplification of force such as when using a screwdriver to open a can of paint.

Levers are measured by Torque. Torque is a Force acting in a direction not in-line with the surface it is encountering ( in this case, the lever arm ).

$$\Gamma = F \times r$$

This answer is a vector quantity, but we want the magnitude of the value, so we can use this form :

$$\Gamma = F*r*\sin(\Theta)$$

In the case of the Lever System, we can simplify this by examining the case of the Force acting in a perpendicular manner to the radius ( on the lever arm ), so that $\sin(\Theta) = \sin(90°) = 1$

Also when the system is in equilibrium, the Torques on both sides are balanced, hence

$$\Sigma\Gamma = 0$$
$$\Gamma 1 = \Gamma 2$$

The radius is the distance along the lever arm, which we will consider as Effort Arm Distance ( $d_E$ ) and Resistance Arm Distance ( $d_R$ )

The equation then becomes :

$$\mathbf{F_E*d_E = F_R*d_R}$$

This idea is parallel to the concept of the amount of work done on or by both sides of the lever. Though the force and displacement are at right angles to each other, the results become the same since each of the lever arms move the same angular distance along which the force is acting and it relates to the radial distance it is from the fulcrum, hence the radius factors into the equation, which is the distance of the lever arm. Recall the concept of Conservation of Energy, where in the ideal case the Input Energy is the same as the Output Energy :

The Energy input = the Energy output.
This becomes
Work Input = Work Output. ( where $W = F*d$ )

$$\mathbf{F_E*d_E = F_R*d_R}$$

This idea is best illustrated with the lever. Returning to the prior example, notice when you use a screwdriver to pry open a paint can lid, or notice where the cutting blade is on a can opener as compared to your hand on the grip as related to the fulcrum in each of these cases. When using a can opener ( a 2nd class lever ) [ note photo ] you exert a given amount of force at a distance that is always greater than the cutting blade wheel is from the fulcrum. Let's say that the ratio of your hand distance to the wheel blade is a factor of 4. This means whatever force you exert is multiplied by that same factor at the wheel blade as exerted on the can!

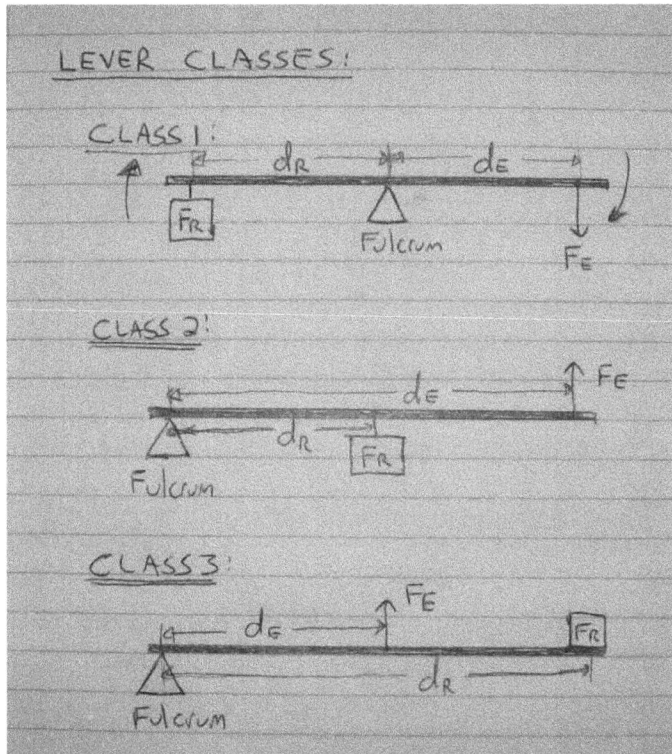

LEVER CLASSES:

CLASS 1:

CLASS 2:

CLASS 3:

Our first thought is that we are getting something for nothing. This is not the case. When calculated the amount of energy put in is all we can get back at most. Recall that Work is the Change of Energy of a system (W = $\Delta$E ). This stems from the general statement of Conservation of Energy, which notes that energy can only be transferred or transformed but cannot be created nor destroyed. Here, when the system is balanced we have a balance of torques or Conservation of Angular Momentum ( which is the rotational version of Newton's 3$^{rd}$ Law ).

We can even measure what is referred to as the Ideal Mechanical Advantage ( IMA ) which is a ratio of the Input distance ( $d_E$ ) to the Resistance distance ( $d_R$ ). This ratio is often greater than 1 for a Class 1 Lever. The best cases are things like a shovel or a jack for a car. Though we have to move the Effort Lever Arm a great distance and the Resistance Lever Arm does not move far, it enables the user to move large masses, like a car.

$$IMA = \frac{d_E}{d_R}$$

Another useful ratio, and in the ideal case is 1 is the Efficiency of a system, in this case the Lever.

$$\text{Efficiency} = \frac{\text{Work}_{Out}}{\text{Work}_{In}}$$

In our Activity, we construct a simple lever out of a classic toy, the Erector set, and use a spring scale ( to measure effort force ), some rulers ( to act as a base for the lever ), and some nuts & bolts ( to act as Load or Resistance Force ) and test the first 2 Classes of Levers and measure the Load, the Effort Force, and the respective distances involved and check to see if there is, indeed, a conservation of energy involved by calculating the work input and output and comparing it to our measurements.

**Purpose :** To compare Work Input to Work Output for 2 different Classes of Levers ( Class 1 and Class 2 ).

**Purpose :** To compare Effort Force to Resistance Force when changing the Effort Arm Distance of a given Lever System with a Resistance Force Load.

**Definitions :**

Classes of Levers :

Class 1 : The Effort and Resistance ( Load ) are on opposite sides of the Fulcrum and move in opposite directions.

Class 2 : The Effort and Resistance ( Load ) are on the same side of the fulcrum so that each moves in the same direction and where the Effort is farthest from the fulcrum.

Class 3 : The Effort and Resistance ( Load ) are on the same side of the fulcrum so that each moves in the same direction and where the Resistance ( Load ) is farthest from the fulcrum.

Other Terms :

Effort Force : The Force ( $F_E$ ) exerted by lever user and measured by the Spring Scale.
Effort Distance : The Distance ( $d_E$ ) from the Fulcrum that the Effort Force is applied to the Lever Arm.
Work Input : The Effort Force times the Effort Distance.
Work Output : The Resistance Force times the Resistance Distance.
Resistance Force : The Load ( $F_R$ ) on the Lever System, composed of a set of Nuts &/or Bolts.
Resistance Distance : The Distance ( $d_R$ ) that the Resistance Force is from the Fulcrum.

## Materials :

- Flat Piece Multi-holed Erector Set Pieces ( either 15 or 25 hole, best ),
- Erector Set Nuts and Bolt,
- 2 Rulers for lever ( Wood & with holes, best ),
- Ruler ( to measure with ),
- Mass Scale,
- Styrofoam Block ( such as from hobby store ),
- ¼" and ½" Nuts & Bolts ( large assortment )( most 5 g to 10 g each ),
- Twist Ties ( small handful ),
- 100g or 250g Spring Scales ( depends on mass used ),
- Slide Rule

## Set-Up of System :

1) In following the directions, look at the accompanying photos below it for illustration.
2) Use 2 wooden rulers with holes and put the bolt from the Erector Set through one of the holes near ( but not at ) the end. Secure it in place with a nut.
3) Now place another nut, the flat piece at a chosen hole to begin with ( best to start with the middle hole ) and another nut in sequence on the bolt.
4) Use another nut and place it far enough along the nut so that another ruler can be now attached and secured with yet another nut to the bottom of the bolt.
5) The end closest to the now constructed lever arm system is treated as an 'A-frame' and can be further secured with a twist tie.
6) The farthest ends of the rulers can now be placed in a Styrofoam block, a small inverted cardboard box, or an inverted Styrofoam plate. Some tape could help this too.
7) Note that you will have to remove the rulers and take off one ruler and one side of the lever arm to move the lever system to another of the holes to be used.
8) Note : Always test the system ahead of operation for the actual Activity so that it performs as expected.
9) Note : A good strategy is to choose the holes to act as the fulcrum ahead of time and make all the measurements for effort arm distances and resistance arm distances.

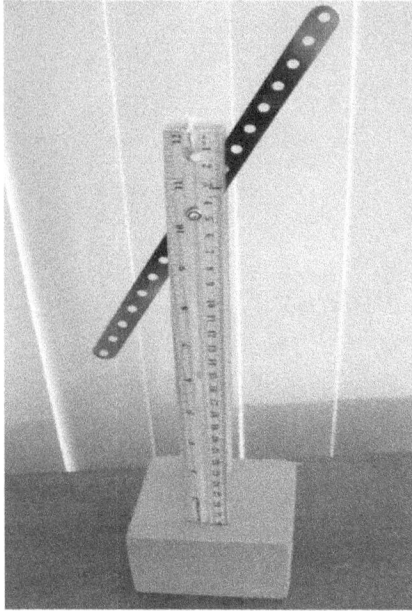

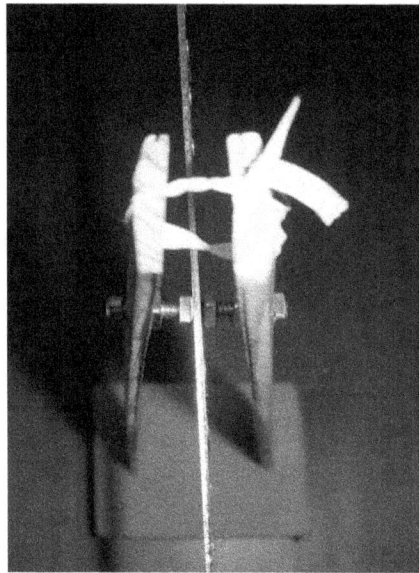

## Procedure :

1) It is best to start with a Class 1 Lever with the fulcrum between the Effort Arm and the Resistance Arm.
2) It is best to start with a given mass ( initial $M_R$ ) and a set resistance distance ( $d_R$ ), for example at the farthest hole from the fulcrum.
3) Note : With regard of mass and the spring scale use find the right combination. It works well with masses in the neighborhood of 25 g to 200 g, at most. Remain consistent with the spring scale used.
4) For the Effort Arm distance ( initial $d_E$ ) start with the spring scale on the farthest open hole on the Effort Arm from the Fulcrum and measure the amount of force ( $F_E$ ) it takes to move the given mass.
5) With each of these trials move the spring scale to a closer hole to the fulcrum and measure the amount of force ( $F_E$ ) it takes. With each trial there will be a new Effort Arm distance ( new $d_E$ ) – this can be measured ahead of time or after the force measurements.
6) In the next set of trials, move the given mass ( $M_R$ ) to a new resistance distance ( new $d_R$ ), such as now at a closer hole to the fulcrum.
7) In a new set of trials, now change the amount of resistance mass by increasing it ( $M_R$ + more material ) and redo steps 2 – 6.
8) Before proceeding with the calculations, be sure to measure the distances from the fulcrum to each : the Effort Arm Force and Resistance Arm Force. When measuring in a given distance, It is best to be consistent with the edge of the hole to measure from ( instead of estimating where the center of the hole is ).
9) When done with these values for a given lever class, try another lever class, such as Class 2, where the fulcrum is at one end, the Resistance Force is between the fulcrum and the Effort Force which is near the other end of the lever. Redo the steps 2 – 8.
10) With a complete set of data, do all the necessary calculations :
11) Calculate the Resistance Force ( $F_R$ ).
12) Calculate the Ideal Effort Force ( $F_E$ ) for each trial.
13) Calculate the Work Input ( $W_{In}$ )( using the Measured Force ) and the Work Output ( $W_{Out}$ ).
14) Graph Work Output vs. Work Input. Draw a best fit line and determine slope. Ideally it should be 1.
15) Use the Percent Error Formula ( % E ) to determine how similar the Ideal Effort Force values are to the Measured Effort Force values.
16) For each trial, calculate the Ideal Mechanical Advantage ( IMA ) and Efficiency ( Eff ).

## Data :

Type of Lever Used : _____

| Trial | Resistance Mass [ $M_R$ ] (g) | Resistance Force [ $F_R$ ] (cm) | Resistance Distance [ $d_R$ ] (cm) | Effort Distance [ $d_E$ ] ( N ) | Measured Effort Force [ $F_E$ ] (N) |
|---|---|---|---|---|---|
| 1 | | | | | |
| 2 | | | | | |
| 3 | | | | | |
| 4 | | | | | |

## Calculations :

Be sure to use your Slide Rule!

## Work :

$$W = F*d$$

## Resistance Force :

$$F_R = m_R*g$$

## Lever Arm Work Relation :

**Work Input = Work Output**

$$F_E*d_E = F_R*d_R$$

**Note : This can be rearranged in a Proportion ( easy on a Slide Rule )**

## Ideal Rearranged Lever-Arm Work Relation :

$$F_E = \frac{m_R*g*d_R}{d_E}$$

## Work Input :

$$W_{IN} = F_E*d_E$$

## Work Output :

$$W_{OUT} = F_R*d_R$$

## Ideal Mechanical Advantage :

$$IMA = \frac{d_E}{d_R}$$

## Efficiency :

$$Efficiency = \frac{Work_{Out}}{Work_{In}}$$

## Percent Error :

$$\%E = \frac{[\ Measured\ Value - Ideal\ Value\ ]}{Ideal\ Value} * 100\%$$

## Needed Constants & Conversions :

$g = 980$ cm/s$^2$ = 9.8 m/s$^2$ ( acceleration due to gravity )
1 m = 100 cm
1 kg = 1000g
$1\ N = 1\ \frac{kg*m}{s^2}$
1 kg = 2.2 lbs.

## Conclusion :

How close are your measurements of Work Input to Work Output? How do your ideal calculations for effort force compare to measured effort force? What sort of ways can this activity be improved and why do the results not match the ideal values? Also, what of the Ideal Mechanical Advantage as well as Efficiency values – in the case of IMA how close to the outcome did it come? What of efficiency – was it near 100%, why or why not?

# Activity #30
## Pulley Exploration Activity
Grade Level : Middle School
Math Level : Calculating


The simple machines are : the pulley, wheel & axle, screw, lever, wedge, and the incline plane. They are a somewhat overlooked item at many levels in all of human technology, from all massive structures to even microscopic-sized-systems today, which are being made to operate as levers, and the like. In our everyday lives, they are found in numerous home tools and applications : scissors, can openers, steps, screws, knives, placement of door handles, all hand tools – pliers, screwdrivers, et al, parts of bikes, parts of cars, parts of engines, and so on.

Remarkably, all simple machines have the same basic outcome. Each of the simple machines can only do one or both of the following : Either change the direction of the force and/or the amount of force. They cannot change the amount of work that needs to be done, but can affect the factors involved in work itself.

One of the most common simple machines ( and being explored in this Activity ) is the Pulley. It is essentially the circular form of a lever. In the case of a lever, it is an 'arm' that rotates about a fulcrum and ,typically, the 'arm' that rotates in the opposite direction is used to first change the direction of the effort force exerted on the first arm and cause a resistance force to be exerted against a load.

The basic Pulley ( one of them ) is used to redirect the force one applies to a string, rope or chain that it goes around and attaches to a mass to be lifted or moved. One has to exert the same amount of force as the weight of the object to be lifted in order to make it move. If so, then why use it? The redirected force may be the answer in this case, but what if there is more than one pulley operating in a coordinated fashion? ( See diagrams )

## Diagrams of Pulley Systems

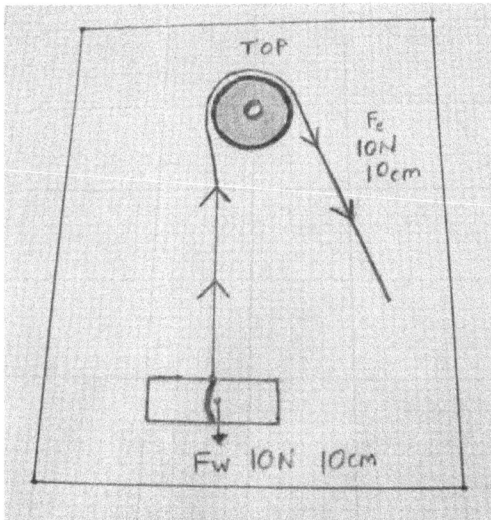

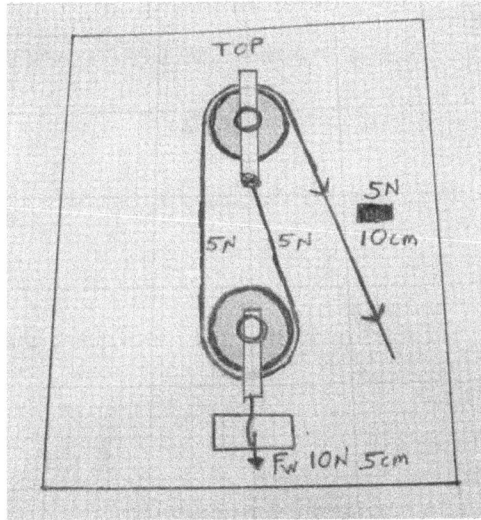

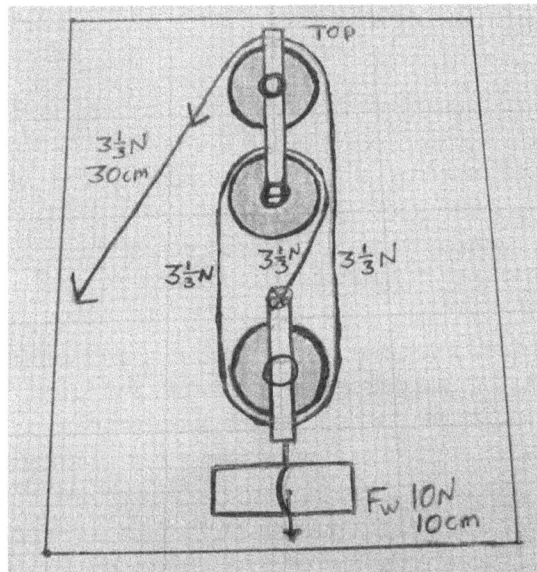

In the case of the Pulley the amount of movement in the direction of the applied force will either be the Input distance ( $d_I$ ) for the Input ( Effort ) Force ( denoted $F_E$ here ) and the Output distance ( $d_O$ ) for the Output Force ( $F_O$ here denoted $F_W$ for Weight but will be $F_O$ in the Activity ).

A force operating through a distance ( displacement technically ) is the definition of Work. In this case, the Effort or Input Force is something or someone pulling the string, chain, or rope in a given direction so as to turn the pulley. The string, rope, or chain is then attached to a mass, that has Weight when moving vertically, so it too is a force, namely the Output Force and it moves a distance as well. Only in the case of a single pulley does the Input and Output Distances equal one another. In the matter of multiple pulley systems, these distances are not equal.

However, what is equal in all cases ( ideally ) is the amount of Work Input will be equal to the Work Output. It is here we begin :

**The Energy input = the Energy output.**

This becomes the equation :

**Work$_{Input}$ = Work$_{output}$**

**$F_E*d_I = F_O*d_O$**

( E = Effort or Input, I = Input, O = Output )

This idea for the Pulley is parallel to the concept of the amount of work done on or by both sides of a lever as well. For further explanation of these ideas contained here ( such as Work, Energy, et al ) read the Preludes to Mechanical Energy Activity, the Lever Activity, the Efficiency Project Activity, Work on an Incline Activity, and Energy & Power on the Stairs Activity.

As noted earlier, in the case of multiple pulley systems, the Input and Output Distances are not equal, yet the product of the given distance with its corresponding Force ( Input or Output ) must equal the other product of the 'complimentary' variables. So, this implies that with different values for distances, there must be different values for the Input and Output Forces as well.

These Pulley Systems not only redirect the force, but they also multiply the force one exerts. With two Pulleys one can readily cut the Effort Force needed to lift a Weight in half. So that a 1000 N object needs only 500 N of effort, for example. Despite the apparent 'free' gift of greater ability, there is a cost. Recall that the Conservation of Energy Idea : The Total Amount of Energy in a Closed System Remains Constant. So here, the amount of Energy put in can only be to the total energy output of the system at most ( and here we are assuming an ideal situation – that is, there is no friction ). Next recall that Energy and Work are related to each other ( formula here ) and Work W = F*d, where the Effort Force is our Input and the Weight is the Output Force in this case. The distance for each, respectively, is the Input distance and the Output distance and is a measure of the distance that the rope moves. From the example given of 500 N of effort to move a 1000 N object means that we must move the rope twice the distance that the object moves. So to lift the object a distance of 10 m, we must move our rope on the Effort side 20 m. The product of each side is then the same. $W_{in} = W_{out}$ or $F_{effort}*d_{input} = F_{output}*d_{output}$, 500N*20m = 10,000J = 1,000 N*10m

We can even measure what is referred to as the Mechanical Advantage ( MA ) which is a ratio of the Input distance ( $d_I$ ) to the Resistance distance ( $d_O$ ). This ratio should be greater than 1 for multiple pulley systems. Though we have to move the string, rope, or chain a greater distance and the output distance is smaller and does not move far, it enables the user to move large masses due to the multiplying effect of the effort force as compared to the output force.

$$MA = \frac{d_I}{d_O}$$

Another useful ratio, and in the ideal case is the Efficiency of a System :

$$Efficiency = \frac{Work_{Out}}{Work_{In}}$$

In our Activity, we use a small set of Pulleys, a tension scale, a small object acting as our mass and measure the amount of force needed to move the given weight and measure the distances our efforts require to move the weight a measured distance. The data then is used to find the amount of Work Input and the Work Output. We can then compare these values to each other and to the idealized expectations. In the process we explore relationships of pulley systems and the advantages of multiple pulleys so as to move larger weights.

## Definitions :

Pulleys : Are wheels with grooved rims that have a string, rope, chain, or belt running through them so as to change the direction of an applied force. – Note : When used in the proper set-up, not only can force be redirected, but also the applied force can be multiplied.

Other Terms :

Effort ( Input ) Force : The Force ( $F_E$ ) exerted by pulley user and measured by the Spring ( Tensions ) Scale.
Effort ( Input ) Distance : The Distance ( $d_I$ ) the string ( rope, chain ) travels while undergoing the Effort ( Input ) Force.
Work Input : The Effort ( Input ) Force times the Effort ( Input ) Distance.
Work Output : The Output Force times the Output Distance.
Output Force : The Load ( $W_O$ ) on the Lever System, composed of a set of Nuts &/or Bolts, battery or other mass used.
Output Distance : The Distance ( $d_O$ ) that the string ( rope, chain ) attached to the output weight moves due to the effort ( input ) force.

**Purpose :** To set up, test, measure, and calculate the Work Input and Work Output of various Pulley system constructions. These measurements are from the determination of a weight of a mass, the effort force to move the mass, the distance the mass moves and the distance the effort force input distance is measured.

**Materials :**

- Tension Scale ( Note : scale range depends on the mass used as the output weight to be moved – typically 100g or 250g best ),
- Mass to be used as weight ( Can be Bolt & Nuts, a Battery – D size or 9V – Other, such as small bag of a few Marbles ),
- Pulleys ( minimally 3 is best ),
- String,
- Scissors,
- Twist Ties,
- Tape,
- Ruler,
- Open-Form Plastic Crate ( see photo ),
- Goggles,
- Slide Rule

**Set-Up of System :**

- It is best to use the illustrations in the Prelude and the photos below to have an idea of how to work with the pulleys and materials you have.
- The key is look carefully at the diagrams and mentally trace the path of the string through it and where it attaches.
- If the diagrams are not enough you might consider some research online to further examine the connections.
- Make the connections for a given system, diagram it for the Data section of the Activity. Be Patient and be certain to test the set-up.

**Photos** of Equipment and a picture of a 2 Pulley Set-Up ( Note : use photos in Prelude for other arrangements )

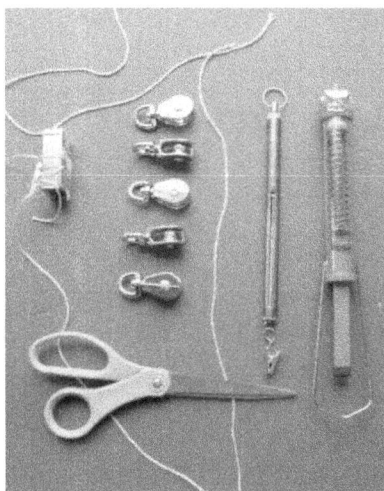

## Procedure :

1. Be sure to zero out the chosen Tension Scale you are using and that it can measure the chosen Mass for your situation.
2. Measure the Mass with your Scale and record it in the Data table. Note that it is labeled 'Weight', so if your Tension Scale is measuring 'grams' use the conversions in the provided conversion table to convert it into 'Newtons'.
3. It is best to wear goggles to maximize safety. Be sure to have parental permission and supervision as well.
4. It is best to start simply ( for example, explore one pulley to understand its operation, how you will measure both the input and output distances and the like ).
5. Decide on a plan of action : Such as 1 Pulley, 2 Pulley, 3 Pulley Investigations. Even in the case of a 2 Pulley or 3 Pulley situation – there may be other ways to arrange them – do the research – conduct some trial and error processes in your investigation, but come up with a plan. The illustrations and the photo should be helpful.
6. Diagram each of your plans and sequentially number them. Use the Data table below for measurements and make as many needed copies for subsequent trials.
7. With each and every Pulley System investigated, it is best to study the diagram first by following the path of the string(s) through the system and decide ( with regards to your equipment ) how to best configure it.
8. For example, 'Twist Ties' are noted on the list – one could simply use string ( as I did ), but some connections between the crate and the pulley or between two pulleys or between the mass and the pulley ( as needed ) by twist ties.
9. It terms of the Pulley Systems under consideration you might consider the following courses of action :
10. First you might conduct multiple trials where you not only hold the output weight constant ( and this should be the case in all of your systems ) but you might want to have an established and set Input Distance – say 4.0 cm ( Note your distances should be measured in the nearest $1/10^{th}$ of a centimeter, but you will need to convert it to meters for the calculations ). In doing so, you can later average these values and see how similar they are. It is best to do at least 3 constant trials when doing this.
11. A second idea is to have a series of incremental changes to the Input Distance ( for example : 4.0 cm, 5.0 cm, 6.0 cm, etc ) and see how this affects the outcome. This can be useful in later calculations and considerations since you might graph this data.
12. The recommendation here is to do both of these things. The calculation directions will assume that this is what you have done.
13. How to Measure Distances :
14. It is easy enough to see how to measure Output Distance – It is merely the amount of distance traveled by the Mass when it is lifted from a starting resting position. Be sure to notice whether your ruler has '0' at the edge or a little ways into the ruler as this may affect the measurement.

15. In the case of Input Distance, you might consider using a small piece of tape on the string you are going to pull. Then place it in a place so that a ruler placed at it and parallel to the bottom of the crate spans across to a spot on the crate ( such as a junction point ) so that this can be considered the zero point. Always try to pull the string as vertically as possible, since angles can affect the distances traversed.
16. The distance can then be determined by using the crate side – the starting point of the tape and the ending point of the tape ( when projected onto the crate  side ). The distance between these two points is the Input Distance.
17. Note that you are not just pulling the string. You actually attach the Tension Scale and pull with it because you have to read the Input Force as well.
18. Be sure to test the system so that it operates effectively before taking measurements.
19. Note all of the Measurements we are to record as they were discussed : Weight of the Mass to be moves ( Output Force ), the Output Distance, the Input Distance, and the Input Force. These are the only needed measurements for our calculations.
20. Calculations :
21. If needed you should have converted your mass measurements into Newtons from the Weight ( Output ) Force measures.
22. Create a Table of Work Input and Work Output. For each, calculate the Work Input and Work Output for each of your trials. If you have done multiple trials of one set-up for a given distance that you kept constant, not only compute these but also calculate an average value for it.
23. Be sure to take averages of constant trials and not the sequential trials where you have incrementally increased the Input Distances. These are their own unique data points.
24. Look at the given data pairs for a system trial for Work Input and Work Output – are they similar or not? Why is this so?
25. Graph the Work Output ( y-axis ) vs. the Work Input ( x-axis ) data points on a graph. Draw a best fit line and determine the slope. Ideally the slope will be 1.
26. In a continuation of the created table, now include columns of Mechanical Advantage and Efficiency. Use the provided formulae to determine these values. Take the average value for a given Multiple Pulley set-up for Efficiency and determine its average. Compare this value to your slope.

**Data :**

Depiction ( & Number ) of Pulley System Used : _____

Your Diagram :

| Trial | Weight [ $W_O$ ] (N) | Resistance Distance [ $d_O$ ] (cm) | Effort Distance [ $d_I$ ] (cm) | Measured Effort Force [ $F_E$ ] ( N ) |
|---|---|---|---|---|
| 1 | | | | |
| 2 | | | | |
| 3 | | | | |
| Average | | | | |

## Calculations :

Be sure to use your Slide Rule!

## Average :

$$X_{ave} = \frac{\Sigma x}{n}$$

( x = a given value in the set, n = the number of elements in the set, $\Sigma$ means the sum of the items )

## Slope :

$$m = \frac{\Delta Y}{\Delta X}$$

## Work :

$$W = F*d$$

## Output Force ( if using a measured mass ) :

$$F_O = m_O*g$$

## Work Relation :

**Work Input = Work Output**

$$F_E*d_I = F_O*d_O$$

**Note : This can be rearranged in a Proportion ( easy on a Slide Rule )**
**Note : For your measurements and calculations, $F_E$ come from the Tension Scale readings**

## Work Input :

$$W_{IN} = F_E*d_I$$

## Work Output :

$$W_{OUT} = F_O*d_O$$

## Mechanical Advantage :

$$MA = \frac{d_I}{d_O}$$

## Efficiency :

$$\text{Efficiency} = \frac{\text{Work}_{Out}}{\text{Work}_{In}}$$

## Needed Constants & Conversions :

$g = 980 \text{ cm/s}^2 = 9.8 \text{ m/s}^2$ ( acceleration due to gravity )
$1 \text{ m} = 100 \text{ cm}$
$1 \text{ kg} = 1{,}000\text{g}$
$1 \text{ N} = 1 \frac{\text{kg}*\text{m}}{\text{s}^2}$
$1 \text{ kg} = 2.2 \text{ lbs.}$

## Conclusion :

How close are your measurements of Work Input to Work Output? How do your ideal calculations for effort force compare to measured effort force? What sort of ways can this activity be improved and why do the results not match the ideal values? Also, what of the Ideal Mechanical Advantage as well as Efficiency values – in the case of IMA how close to the outcome did it come? What of efficiency – was it near 100%, why or why not? Finally, how do the various pulley systems that you have tested work out? As you increased the number of pulleys, did your effort force increase, decrease, or stay the same? Why? What advantages do multiple pulley systems have over ones with few or one pulley in them?

# Activity #31
## The Measure of Work on an Incline ( Ramp ) Activity
Grade Level : High School
Math Level : Calculating


Work, by definition, is the change of energy of a system. Also, from the relation for Work, it is a Force acting through a Displacement ( $W = F*d*\cos\Theta$ ). The angle factors in when the Force and the displacement are not in the same direction and this factor takes the component of the Force that acts in the direction of the displacement.

This formula leads to many famous questions concerning work, such as : "How much work is done when holding a tray from beneath and walking a given distance on a level floor?". Strangely, the answer is zero! This is because the Force is acting up ( vertical ) and the displacement is horizontal. The $\cos(90^\circ)$ is zero.

When we lift an object manually or by machine in a vertical fashion we are changing its gravitational potential energy. We are increasing it as it is moved vertically upward. Interestingly we could lift a box of books onto a high shelf above our heads with our hands and arms, use a pulley to do this, push it up a ramp by sliding the box on the ramp ( and neglecting friction ), or even have a machine pull a rope to drag it up the ramp, and in each case the amount of work is the same!

Why then use a ramp? First, what is a ramp? It is an incline to any horizontal surface that acts in a vertical direction. It is always a longer path than going straight up vertically, as it is the hypotenuse of a triangle between the horizontal and vertical directions under consideration. But like all simple machines ( remember those? The pulley, wheel & axle, screw, lever, wedge, and the incline plane ).

**Each of these can do one or both of the following : Either change the direction of the force and/or the amount of force. They cannot change the amount of work that needs to be done, but can affect the factors involved in work itself.**

What then is the ramp used for? By increasing the distance traveled, the amount of force decreases proportionately. This is due to the conservation of energy :

The Energy input = the Energy output.
This becomes
Work Input = Work Output.
$F_I*d_I = F_O*d_O$

This idea is best illustrated with the lever ( instead of the ramp ). Notice when you use a screwdriver to pry open a paint can lid, or notice where the cutting blade is on a can opener as compared to your hand on the grip as related to the fulcrum in each of these cases. When using a can opener ( a $2^{nd}$ class lever ) you exert a given amount of force at a distance that is always greater than the cutting blade wheel is from the fulcrum. Let's say that the ratio of your hand distance to the wheel blade is a factor of 4. This means whatever force you exert is multiplied by that same factor at the wheel blade as exerted on the can!

Our first thought is that we are getting something for nothing. This is not the case. When calculated the amount of energy put in is all we can get back at most. Recall the Conservation of Energy rules : Energy cannot be created nor destroyed ( in a closed system ) it can only be transferred and/or transformed.

What about the case of the ramp? Notice that if the work in moving an object up a ramp is its change in gravitational potential energy, then we have :

$W = \Delta E$
$F*d = m*g*h$

Since the gravitational Potential Energy is constant, if displacement is changed, say it is larger, then the Force needed is smaller by the inverse of that factor! It also stands to reason that the amount of energy we input is the only energy that the system can have and that which we can get back.

Of course, we are assuming no loss of energy due to friction, which means our input would have to be greater than the output. Realize that the output will at most only be equal to the input and almost always less but never greater. This leads to efficiency :

$$Eff = \frac{Useful\ Work\ Output}{Total\ Work\ Input}$$

Our goal of this activity is simple, then. Construct a small ramp and move a mass along it and measure the amount of work done to elevate it to a given height. Compare this value to other ramps of differing angles, hence lengths to examine the amount of force needed to accomplish the task and comparing the amount of work done in each case. We should come to find that independent of the path, the work is the same.

**Purpose :** To examine the changes in force and distance when using an incline at different angles and their relation to the amount of work done.

## Materials :

- Board ( 2-3 ft long, about 6 in. wide at most ),
- Block of wood with a hanger hook on end that is attached to the block,
- Stack of Books or a plastic milk crate,
- Tension scale ( measures in Newtons ),
- Measuring Tape,
- Protractor,
- Slide Rule

## Procedure :

1) On a level table place a milk crate standing on end ( or stack of books ) so that the board that acts as the distance to travel for the block of wood leans against it and can do so at various shallow and steep angles. Note a book can hold it in place.
2) Use a pencil to mark a starting point for the board about 2-3 inches from the end nearest the table. One edge of the block will start here each time.
3) Set the system up so that the angle of the board with respect to the table is 60°. Use the protractor for each measured angle.
4) The height of the milk crate is held constant and measured ( h ).
5) The distance that the block moves is to where the edge of the crate meets the board ( d ).
6) The goal is to have the block only move the same vertical height ( h ) in each trial independent of angle, but with each angle of the ramp, the total distance covered will vary ( d ).
7) Note that only the board will now move, going from the maximum to the minimum angle. As the board is moved back away from the crate in each trial, the distance traveled increases for the block.
8) Be sure to have zeroed out the tension scale before use.
9) Start with the block at the starting point and pull with a constant force for a distance that reaches where the block has met where the board meets the top of the milk crate. ( Note : As you change the angle, the distance that the block needs to travel will change as well ! )

10) Mark down the force ( F ) measurement, and measure the distance ( d ) and record it as well.
11) Conduct each trial 3 times and average the forces ( $F_{ave}$ ) and distances for the final table of results to be used for calculation purposes.
12) For each trial Calculate the amount of Work that is done and compare these values to each other. (This is from the Work on Ramp Formula ).
13) Also compute the Ideal Work on Ramp formula values using Gravitational Energy formula to compare the amount of work you did to the ideal value.
14) For further investigation, decrease the friction as much as possible for the system ( cover each in aluminum foil for example ) and redo the experiment and compare the work done to the change in energy ( gravitational potential energy ) of the system.
15) Further analysis can come from first determining the coefficient of sliding friction ( see activity for this ) of the block on the board and us this to determine the predicted amount of work and compare this total amount of work to the actual work done in each case.

## Data :

Mass of Block : _____ kg

Weight of Block : _____ N

Height ( h ) : _____ m

| Trial Angle | Force [$F_{ave}$] ( N ) | Distance [d] ( m ) |
|---|---|---|
| 15° | | |
| 30° | | |
| 45° | | |
| 60° | | |

## Calculations :

Be sure to use a Slide Rule for all calculations!

Formulae used to find formula needed :

**$F = m*a = m*g$**   ( form depends on where used )

**$W = F*d*cos\Theta$**
                ( here $\Theta = 0°$, so $cos(0°) = 1$ )

$W = \Delta E$   ( note that we are using PE and $PE_i = 0$, $W_i = 0$ )

<u>Calculated Formula for Work on Ramp :</u>

**W = F*d**

Formula for relation of 'h' and 'd' :

$$\textbf{Sin}(\Theta) = \frac{\textbf{h}}{\textbf{d}}$$

<u>Final Ideal Formula ( i.e. no friction ) :</u>

**PE$_g$ = m*g*h**
                    g = 9.8 m/s/s

Note : This formula is for comparison to your measured Work !

<u>Further exploration formula :</u>

$F_f = \mu * F_N$

Predicted Total Work = Change in Gravitational Potential Energy +
                        Work Done Against Friction

Predicted Total Work : **W = m*g*h + $\mu$*m*g*d**

Note : Predicted work can only be done if the coefficient of
sliding friction activity is done first so that the coefficient of sliding friction
is available. – Also 'h' here is the height of the ramp and not the distance
traveled by the block ( which is 'd' )!

Note : Also, if there was no friction ( i.e. $\mu = 0$ ), then the only Work we
would need to do is lifting the mass the given height !

## Conclusion :

The idea in this activity is to compare the amount of work in each trial. Are they very close to each other or not? Also, if you have taken the activity further and either factored in the coefficient of sliding friction or reduced it so much that it is not critical, then how does this amount of work compare to the change in gravitational potential energy?

Notice too that the Ramp ( a simple machine ) being used here did NOT change the amount of work, only the amount of effort! But at what cost? The question is this : What happened to the amount of force needed to move the block as the angle was changed? Did the force increase as the angle increased or vice versa? With further study of the activity and/or the formula can you determine the answer? What of the distance traveled? As force increased, what happened to distance?

Below are some of the answers to the ideas are explored.

The key to this activity is the realization that independent of the path, the amount of work should be the same! This also means that since Work ( W ) does not change, yet the amount of Force changed in each of the trials, so to must the distance change only in the opposite direction. For example, if force increased, then the distance decreased. What happened to the angle? For the force to be greater, the angle is larger, or steeper. Anyone who has used a ramp understands this. The longer the ramp, the lower the effort force needed to move the object. Unfortunately the lower force is at the expense of a larger distance to travel.

# Activity #32
## Work due to Friction versus the Mechanical Energy of a Ramp
Grade Level : High School
Math Level : Calculating

Read about the concept of Energy in the other Mechanical Energy Activity on the website www.cosmicquestthinker.com or this same Activity is found in volume one of The Inquisitive Pioneer. This is an alternative Activity and involves not only Gravitational Potential Energy and Kinetic Energy but also their connection to the concept of Work as well. Some basic ideas needed here are ideas like Conservation of Mechanical Energy – that is the energy we begin with is what we end with. The Gravitational Potential Energy is first turned to the Kinetic Energy of motion of the ball and then the frictional force acting on the ball does work to bring the kinetic energy to zero – all of the energy has been converted to heat, sound, and some small amount of distortion of the surface the ball rolls on. Also there is another Energy Activity in this book involving Elastic Potential Energy. There is also an Activity involving Work and the Incline.

**Purpose :** To use the conservation of energy to measure the amount of work done by a rolling ball being stopped on a surface with a lot of friction.

**Materials :**

- Stack of Books,
- Dowel Rods,
- Meter Stick,
- Marble ( large is a good choice ),
- Carpeted floor or rug(s) / towel(s),
- Measuring Tape,
- Timer,
- Slide Rule

**Pre-Activity Set Up :**

1) In a carpeted room make a stack of books from which the dowel rods are placed with a small enough gap so that a marble ( large is best ) can roll down it as a track. Keep the angle shallow ( i.e. 2-4 books at most ). Keep the track only as long as a meter stick ( see set up point #2 ).

2) Parallel to the track and as long as the track set a meter stick so that the zero point is at the end where the marble would roll off.

3) The best case has the ends of the dowel rods and meter stick at the edge of the top of the book stack with just enough to make it stable. Tape will help. Also have these items near the edge of the books since the height of the stack and the base length of the triangle made by the rods will be measured and used here.

4) Be sure to have enough carpeting for the ball to travel on ( 3-8 m – depending on thickness ). The test is when the ball is at the highest point on the ramp and released, does it stay on and stop on the carpeting, if so then okay.

5) Note : that the distances may be too far for rugs and/or towels being used in place of not having carpeting, so in that case shorten the track run starting points and be sure to use a shallow angle so that it does not develop a large speed.

**Procedure :**

1) Measure the mass of the ball.
2) Measure both the height and base length of the ramp using the measuring tape.
3) Use the starting positions noted in the table and begin the marble at those positions ( c ).
4) Let the marble go, and at the moment of start, begin the timer and measure the time it spends on the track system ( t ).
5) The marble now leaves the track and slows on the carpeting.
6) Measure with the measuring tape the stopping distance of the ball from the end of the track to its final position ( s ).
7) For each starting position, do this 3 times and average the values for time and stopping distance.
8) Repeat steps 3-7 until all of the positions are done.
9) Use the data to calculate the angle of the track.
10) Determine the height of the track for each of the starting positions ( h ) from the angle determined.
11) Calculate the Predicted Speed and the Measured Speed and complete the table below.
12) Calculate the Potential Energy, the Kinetic Energy, and the average Force and place these values in the other table below.
13) Graph the stopping distance ( save ) vs. the starting height ( h ) [ Note in all Activities, the y-axis is noted first, then the x-axis ]
14) Draw the graph of Kinetic Energy ( KE ) vs. the starting height ( h ).
15) For each graph draw a best fit line and calculate the slope.

## Data :

Mass of the Ball [ m ] : _____ ( g )

Height of Ramp [ H ] : _____ ( cm )

Base of Ramp [ D ] : _____ ( cm )

| Trial | [ h ] calc. height (cm) | [ c ] start position (cm) | $(s_1)$ stop distance (cm) | $(s_2)$ stop distance (cm) | $(s_3)$ stop distance (cm) | $[s_{ave}]$ stop distance (cm) |
|---|---|---|---|---|---|---|
| 1 | | 30 | | | | |
| 2 | | 40 | | | | |
| 3 | | 50 | | | | |
| 4 | | 60 | | | | |
| 5 | | 70 | | | | |

| Trial | [ h ] calc. height (cm) | [ c ] start position (cm) | $(t_1)$ time of roll (s) | $(t_2)$ time of roll (s) | $(t_3)$ time of roll (s) | $[t_{ave}]$ average roll time (s) |
|---|---|---|---|---|---|---|
| 1 | | 30 | | | | |
| 2 | | 40 | | | | |
| 3 | | 50 | | | | |
| 4 | | 60 | | | | |
| 5 | | 70 | | | | |

## Calculations :

Be sure to use your Slide Rule!

| Trial | [v_p] Predicted Speed (m/s) | [v] Measured Speed (m/s) |
|---|---|---|
| 1 | | |
| 2 | | |
| 3 | | |
| 4 | | |
| 5 | | |

| Trial | [h] starting height ( m ) | [PE] Potential Energy ( J ) | [v] Measured Speed (m/s) | [KE] Kinetic Energy ( J ) | [s] Stopping Distance ( m ) | [F] Average Force ( N ) |
|---|---|---|---|---|---|---|
| 1 | | | | | | |
| 2 | | | | | | |
| 3 | | | | | | |
| 4 | | | | | | |
| 5 | | | | | | |

Be sure to convert centimeters to meters and grams to kilograms to have regular units in Physics ( 1000 g = 1 kg ) ( 1 m = 100 cm )

For calculations, use g = 9.8 m/s/s

$$\tan\Theta = \frac{H}{D}$$

$$\Theta = \tan^{-1}\left(\frac{H}{D}\right)$$

Predicted Acceleration : $a = \frac{5}{7}*g*\sin\Theta$

Predicted Speed : $v_p = a*t$ ( used measured 't' here )

Measured Speed : $v = \frac{2*c}{t}$

'c' is the starting point of the ball's distance from ramp bottom end

Indirect Calculation of Ramp Height for Starting Point :

$$h = c*\sin\Theta$$

$$PE = m*g*h$$

$$KE = \frac{1}{2}*m*v^2$$

Work Relations :

$$W = F*d$$

$$W = \Delta KE$$

$$F = \frac{\Delta KE}{d}$$

Note : d=s

Conclusion :

How do Measured Speed and Predicted Speed compare?
With increasing height, what happens to the kinetic energy of the ball?
What sort of line and what is the slope of the line for the Kinetic Energy ( KE ) and Starting Height ( h ) Graph?
What sort of line and what is the slope of the line for the Stopping Distance ( s ) and Starting Height ( h ) Graph?
Since the ball came to a stop, yet had initial Kinetic Energy, what became of this energy? ( Look at the Work-Kinetic Energy formula and theorem ).
Assuming a constant average force, what was the average force ( F ) affecting the ball over the distance of stopping?
How does the coefficient of ( rolling ) friction affect this outcome ( that is to say, what if we used a thicker carpeting )?

Summary :
Extended Idea :
In this Activity, instead of a marble, one could use a golf ball, ping pong ball, tennis ball, or a wood or steel ball. These could be used for comparative purposes. ( Notes : First, the Predicted Acceleration formula is for a solid ball, so will have some variance for some of these in the list. Second, some may not work so well, why )
Also : Consider this question : What factors affect the stopping distance of the ball and if compared to cars, what would these factors be? ( initial speed and the force of friction is the answer ).

# Activity # 33
## Conservation of Momentum in a Reverse Inelastic Collision Activity
Grade Level : High School
Math Level : Calculating

Momentum is the product of mass and velocity. Like energy and more universally than it, it is always conserved in all situations ( elastic and inelastic collisions ). Newton's $3^{rd}$ Law can be used as an extension or argument for the conservation of momentum. Basically, Newton's $3^{rd}$ Law states 'for every action force on a given mass, there is a corresponding reaction force on the mass imparting the action force that is equal in magnitude to the action force, but opposite in direction. This results in the idea that when two masses collide or when a given mass separates into pieces, the total of the momentum of that situation remains constant. The main ideas of conservation that are contained in this Activity are discussed in the Activities concerning Mechanical Energy ( found online at www.cosmicquestthinker.com as well as in the first volume of The Inquisitive Pioneer ) and in this activity concerning the Conservation of Momentum. Basically they state that the amount that one starts with in a closed system ( energy or momentum ) one ends with in a given situation. Here the foundations of Science and Physics are noted, where the Conservation of Energy and Momentum are outlined.

In the case of an Elastic Collision ( where the impacting particles remain intact and are not conjoined ), both the Momentum and Energy are Conserved. This means that the initial amount before the impact is equal to the outcome value after the collision.

As noted in the Preludes, Energy, like Momentum, cannot be created nor destroyed. The amount that a system has going into an interaction is the maximum amount that can come out of it.

### $Energy_{Before} = Energy_{After}$

### $Momentum_{Before} = Momentum_{After}$

Energy is not conserved in Inelastic Collisions because some of the Kinetic Energy turns into heat, sound, and other forms of energy in the process. In both Inelastic and Elastic Collisions, however, Momentum is conserved.

– for further reading, look at the Activity for Newton's $2^{nd}$ Law ( in the first volume of The Inquisitive Pioneer ) and the Lever Activity ( Activity #  ) as well as the Mechanical Energy Activity ( already noted previously ) for concepts of conservation of various quantities. In this Activity, we will examine the idea of a system coming apart – where two masses, two roller skates, are pushed apart by magnets that have the same pole facing each other so that they repel. By measuring their respective masses and their distances and times ( to compute average speed ) we can then find their momentum values and compare them to each other. There are many examples used in everyday life to illustrate the notion of conservation of momentum as represented here in this case. We could consider an atom breaking apart from radioactive decay. The sum of the momentum of the particles racing off, even at angles to each other and a given system, will add up in a vector fashion ( momentum is a vector, but here we are only concerned with its magnitude and not its direction ) to zero. A firing cannon or gun or even a person throwing something illustrates this same phenomena. In the case of the cannon, the lower mass cannon ball projectile flies in one direction at a high rate of speed, while the cannon recoils at a much slower speed   ( since it is more massive ) in the opposite direction. Looking at this list, one might say, wait a minute – I throw things often ( baseballs, Frisbees, et al ) yet I do not go flying off in the opposite direction to it. This is because of not only you mass as compared to the object you are throwing, but also the frictional force between you and the ground. A fun thought experiment question poses the idea of you standing at the center of a frictionless ice pond with a massive ball in your hands ( why this is so, who knows – it's an imaginary thing ). The question is how do you get off the ice field? Simple – throw the ball in a given direction. You will move, albeit slowly, in the opposite direction at a constant speed ( more Newton's Laws stuff ) until you reach the edge of the ice pond.

**Purpose :** To have two roller skates be initially at rest, then separate due to an impulse imparted to both and determine the momentum of each to demonstrate the conservation of momentum.

## Materials :

- Pair of Roller Skates
- Note : if available use a Collision Cart Set ( one with spring plunger ),
- Either long flat table or tiled floor ( on the floor is best ),
- 2 Timers,
- 2 Meter Sticks or Measuring Tape,
- Set of strong Ceramic Magnets,
- Mass scale ( capable of measuring the roller skates ),
- Masses ( boxes of metal nuts, bag of marbles, other ),
- Sandwich bags,
- Tape or String,
- Scissors,
- Slide Rule

## Pre-Activity Set Up for activities:

1) For this set of Activities, there needs to be a region of about 2-2.5m in length for the roller skates. In most cases, the floor is the best choice over tables, since we need to stop the skates moving in two different directions and times.

2) Mass is needed to be known, so take the mass of both skates individually and record this in the data tables provided ( M1 & M2 ).

3) With the Ceramic Magnets determine which two sides of the magnets are the same – that is, they repel ( can be either two Norths or two Souths ) and attach them to the back of the skates so that if the skates are facing so that the heels are aligned and the toes of the skates point in opposite directions, the magnets are now facing each other at the back side of the skates. If pushed together and let go, they will repel since the same pole of the magnet is facing each other. Note : You need strong magnets and well-lubricated wheels so that the skates move when on the tiled floor.

4) You need known masses to add, it is best to use sandwich bags and either marbles or metal nuts. all the better. Make each of the bags to use is about 100g+ as measured on the scale ( so that they effect the overall mass of the skate with some measure ( so that when all added to one of the skates it is up to 2-3 times the original mass. ) You can have up to 4 of these bags that reaches this total.

5) Test out the skates, making sure the wheels move freely.

6) Note that human reaction time will be part of the reason for measurement reliability and accuracy. It is best to have a starting distance for the system about 2 cm in front of the skates ( when facing in opposition ) and time them for a travel distance of something between 10-20 cm total travel ( as long as it appears to be moving at a constant rate of speed ).

7) The way to test the system is to have both skates aligned back to back and toes pointed away from each other. To hold them together, use tape or string so that there is an outward force pushing on them. Keep them as close as possible so that it increased the force. For each trial, however, always maintain a constant separation distance.

8) With the system ready to go ( attached and repelling ) cut the tape or string and watch how they role. You may need to have meter sticks on the floor along the path not only for distance, but for guidance, so that they roll in a straight line. Notice how far they roll, how long each of them takes to roll this distance. Recognize you need to time both of them, so having two people, one to watch each of the skates is best, but this can be done by one person   ( though with greater margin of error ).

9) Be sure to test the system repeatedly so as to have an idea how it will work and to have an idea of how long it takes to travel the distances involved.

## Procedure for Activity :

1) From the Pre-Activity Set Up, you have a good idea of how the system will operate. Now is the time to do the same sort of motion of the skates, but now take measurements.
2) In the data table, you should have the masses of the skates recorded, Be sure to keep track of which is M1 and which is M2.
3) Have an idea of the distance you will use for travel. It may be best to choose a consistent distance for each of the skates, say 10, 12, or 15 cm. This distance is (D1). This way any differences that are present will show up in the time they take to travel.
4) Set up the skates back to back, bound by the string or tape, then cut and let them go.
5) At their start mark, start each of the timers and then stop the timers when they have traveled the prescribed distance. Record the time ( t ).
6) Do this 3 times for a given trial and calculate an average time using a slide rule for this trial ( $t_{ave}$ ).
7) From distance and average time measures, determine the average speed ( $v_{ave}$ ) of each of the skates.
8) Now compute with slide rule the momentum of each of the skates for comparison. The momentum for the skate rolling to the right should equal the momentum of the skate rolling to the left! How similar or different are they?
9) In each of the subsequent trials, now add mass to one of the skates. A more massive skate should not travel as far nor as fast as it did originally, so be sure to pretest this set up to have an idea. You may have to change the distance originally chosen. Redo the 3 trials with this change to mass.
10) In each of the subsequent trials from the previous step, increase the mass up to the maximum mass.

## Data :

| Trial | Cart 1 [M1] (kg) | Cart 1 Time [t] (s) | Cart 1 Distance [$D_1$] (m) | Cart 2 [M2] (kg) | Cart 2 Time [t] (s) | Cart 2 Distance [$D_2$] (m) |
|---|---|---|---|---|---|---|
| 1 | | | | | | |
| 2 | | | | | | |
| 3 | | | | | | |

## Calculations :

Be sure to use your Slide Rule!

### Average Time :

$$t_{ave} = \frac{\Sigma t}{n}$$

   't' is the time for a given trial, 'n' is the number of times

### Average Speed :

$$v = \frac{d}{t_{ave}}$$

### Momentum :

$$p = m*v$$

### Conservation of Momentum :

$$\Sigma \ = 0$$

$$p_1 = p_2$$

## Conclusion :

The examination of the data should show that the momentum of the System, which is the momentum of the skates should be equal. Notice that the speed of a more massive skate should be smaller than the lower-mass skate so that the product of mass and speed balances out. How can you account for the differences in their values?

# Project
## Personal Slide Rule Template

MAKE A PAPER 6" SLIDE RULE

On the following two pages are two different templates than enable you to make a 6" slide rule.

Make your choice as to which you want to construct. They each have the same number and type of scales and are quite complete and useful. They have these scales for use : C, D, C1, D1, CF, DF, C1F, A, B, S, T, ST, K, L

Things Needed : One of the Templates, Scissors, Ruler

Steps to making a slide rule :

1. In either case make 2 copies of the chosen Template.
2. For both you need something to act as a cursor – best choice is a ruler. Be sure to align it with a straight edge, such as the bottom of the paper in the case of your unfolded slide rule and be sure to fold along a straight line so as to be able to use this for the folded model.
3. In the case of the first slide rule which is the unfolded slide – this is the set of scales where they are all bunched together and in separate boxes do the following :
4. Leave the first copy alone. It acts as the stators for your slide rule and simply lie on the table
5. With the second copy, cut out the slide – the middle set of scales – so that it can be moved along between the stators as needed.
6. It is now ready to use – Use a ruler where it lays across the stators and slide perpendicular to the direction one regularly reads the paper and the bottom edge of the rule is aligned with the bottom edge of the paper ( needs to be perpendicular ).
7. In the case of the second slide rule which is the folded slide – this the set of scales in sets of 5 in boxes which are separated.
8. As in the first slide rule, make two copies of the template to be used.
9. With the first copy fold it so that the top and bottom set of scales are now opposite the middle set of scales. It is best to fold it so that it would align with the middle set of scales as if it were a slide and the top and bottom are the stators. – Note if there is excess paper above and below the stators so that it would interfere with the slide and their reading, cut this away in a straight a manner as possible ( follow the line of the box encasing them ).
10. Now fold the second template so that it fits into the sleeve and the middle set of scales faces out of the space between the top and bottom stators and is now the slide.
11. It may take some adjusting, so be patient.

12. Be sure to follow the lines of the boxes for folds as it is critical that the upper and lower stator have aligned scales.
13. With your cursor ( best choice is probably a ruler as in the first case ) be sure that the end edge aligns with the lines and the ruler itself is the cursor line.

Some notes for Use :
In order to use the A/B or CF/DF scales these are on the adjacent slide and stator in a doubled fashion.
Have fun and enjoy : )
Thanks :
These scales came from the web site : The International Slide Rule Museum ( sliderulemuseum.com ) found on the Slide Rule Reference Scales tab and set with graphics by Andrew Nikitin

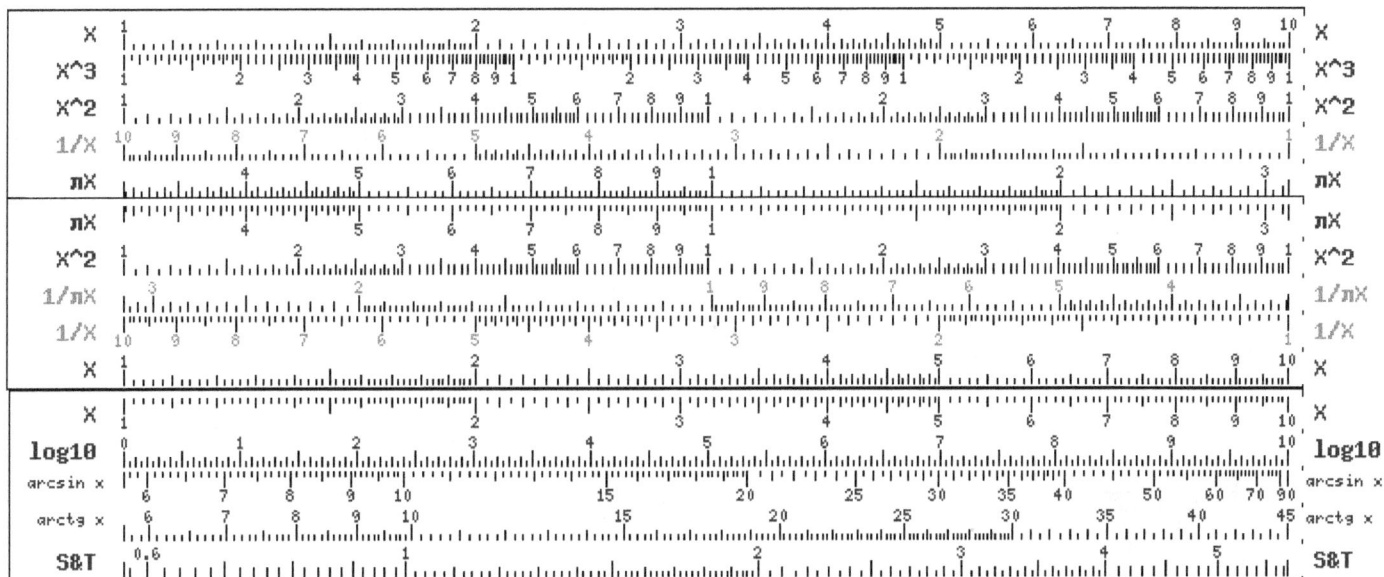

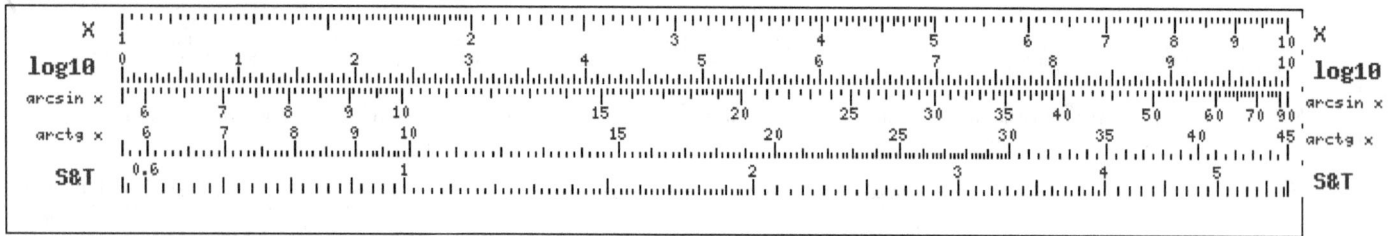

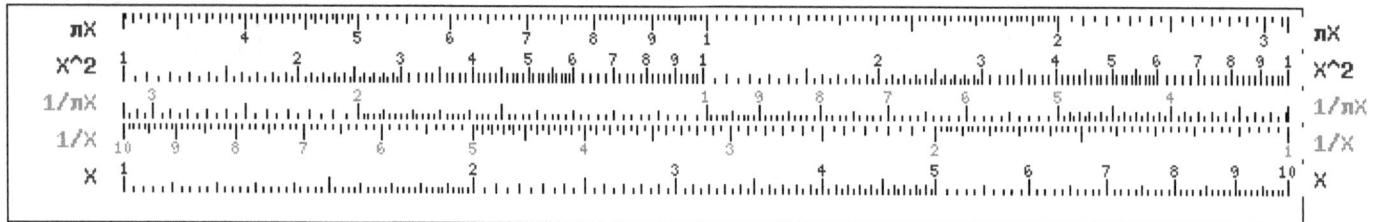

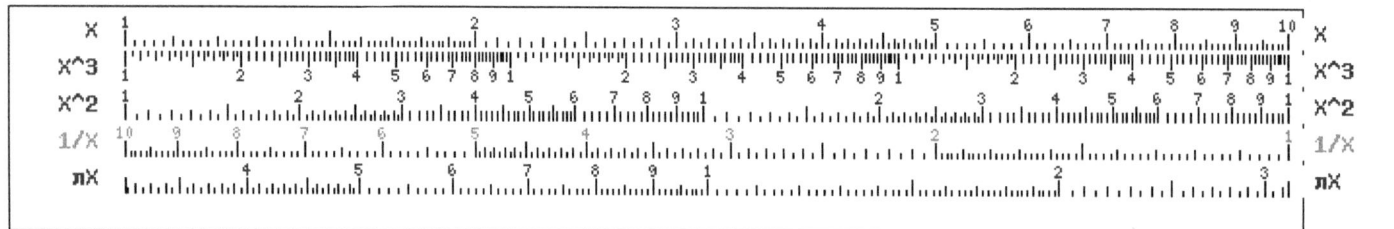

316

www.ingramcontent.com/pod-product-compliance
Lightning Source LLC
Chambersburg PA
CBHW081053220326
41598CB00038B/7073